Simmer or Sizzle

COOKING WITH YOUR

Slow Cooker or Contact Grill

SiMMer or SiZZle

COOKING WITH YOUR

Slow Cooker or Contact Grill

New Lenox
Public Library District
120 Veterans Parkway
New Lenox, Illinois 60451

Kathryn Moore • Roxanne Wyss

THOMSON

DELMAR LEARNING

Australia Brazil Canada Mexico Singapore Spain United Kingdom United States

THOMSON

DELMAR LEARNING

Simmer or Sizzle: Cooking with Your Slow Cooker or Contact Grill
by Kathryn Moore and Roxanne Wyss

Vice President, Career Education Strategic Business Unit:
Dawn Gerrain

Acquisitions Editor:
Matthew Hart

Product Manager:
Patricia M. Osborn

Director of Production:
Wendy A. Troeger

Project Manager:
Ken Karges

Director of Marketing:
Wendy E. Mapstone

Channel Manager:
Kristin McNary

Marketing Coordinator:
Scott Chrysler

Cover Design:
Joe Villanova

Library of Congress Cataloging-in-Publication Data

Moore, Kathryn.
  Simmer or sizzle : cooking with your slow cooker or contact grill/ Kathryn Moore, Roxanne Wyss.
      p. cm.
  ISBN-13: 978-1-4180-3809-0
  ISBN-10: 1-4180-3809-1
  1. Electric cookery, Slow.
  2. Indoor electric grills.
  I. Wyss, Roxanne.
  II. Title.
  TX827.M66 2006
  641.5'884—dc22
                          2006023892

NOTICE TO THE READER

*To our families—Bob and Grace, David, Laura, and Amanda,
who make our lives richer, more meaningful,
and make dinnertimes such cherished moments.*

Contents

Preface

Simmer or Sizzle

Simmer or Sizzle is a cookbook to help you plan and prepare great meals. We combine recipes for both slow cookers and contact grills. These are quick, time-saving meals that every cook, no matter the skill or experience in the kitchen, can make to enjoy the tantalizing flavor that comes from slowly simmered or quick-sizzled meals.

We bring you this cookbook with years of experience from several small appliance and food companies. Over the years we have shared our expertise with their consumers, and now we want to share them with you. Slow cookers and contact grills have been the tools of our trade for years. We were among the very first, and remain foremost, authorities on using the slow cooker. Most of all, we share the "cooking trenches" with every family on the block. We are recognized for great family meals and now are eager to share our professional background and training as home economists, our experience with small appliances, and our tips for family meals with you.

Electric slow cookers are one of the most important small appliances. Their popularity, after more than 30 years on the market, invites requests for great, new recipes. The contact grill is known for its speed and flavor, yet no cookbook has combined these two appliances into one alluring collection—encouraging you to enjoy the comforting simmering one night and the crunchy sizzle the next.

How to Use This Book?

In the introduction, we share our secrets for meal planning—no, not kitchen magic, but ideas you can use to get dinner on the table night after night. In fact, the entire book is packed with great tips you can use no matter what cooking method you might choose that night—tips that will ensure that dinner is on the table quickly and that will make the work a bit easier. Part I tempts you with an incredible collection just for the slow cooker—the simmered specialties include soups, stews, chilies, roasts, and so many more family favorite dishes. Keep reading and you will discover wonderful dips and snacks, side dishes, risottos, desserts, and baking that you never thought possible from a slow cooker. Part II sizzles with juicy grilled steaks, chicken, fish, and other main dishes, crisp sandwiches and panini, and grilled vegetables that shout flavor, in spite of their speed.

So, we invite you to share dinner with us. Enjoy a bit of simmer or sizzle. Most of all, enjoy dinner with your family!

Acknowledgments

We feel so blessed! Each day of our personal and professional journey we are touched by so many wonderful folks who share their love, their skills, their love of food and cooking, and truly the best of themselves with us. We thank them all, yet, space demands we note just a few.

- Roxanne owes special thanks to her parents, who instill love of family and wonderful Sunday dinners her mom has prepared year after year; and to Bob and Grace, thanks for your love and support. You are my perfect recipe for family!

- Kathy owes special thanks to her family—David, Laura, and Amanda, for love and support, day in and day out, and to her parents for supporting dreams, establishing traditions, and making the kitchen the heart of the home.

- We are blessed to work as friends. After so many years of working side by side in kitchens throughout the country, we are truly cherished friends, and that makes the journey so much more rewarding.

- Matt Hart, Patricia Osborn, and the other wonderful people at Thomson Delmar Learning who believed in us, had a vision for this book, and made the book a reality.

- Mr. Conrad Hock and the great folks at Williams Foods, Inc., who create fantastic products and allow us to proudly represent them on so many television shows and at so many events, cooking contests, chili cookoffs, and more for these many years.

- Bill Endres, and each and every person on the team at Select Brands, Inc., for allowing us to work with their great appliances, be a part of their close-knit group, and have fun at the same time.

- Jill Silva, food editor of *The Kansas City Star*, for food inspiration, friendship, and the opportunity to develop recipes for the "Eating for Life" column.

SIMMER OR SIZZLE— THE ANSWER TO "WHAT'S FOR DINNER?"

What's for dinner? Can you almost taste it? That sizzling, simmering allure of dinner calls the family to the table.

Are we dreaming? Those three little words—"what's for dinner?"—usually bring groans and complaints. Not the same old stuff! Not another burger-in-a-bag from a drive-through! No, we understand! It's a nightly challenge. Boredom is pushing chairs away from the dinner tables of too many families. Are you begging for something new, yet realistic, to make? It is time to take charge!

Yet, how can one possibly be successful in the nightly dinnertime battle? Here's the plan. Equip yourself with this collection of recipes designed with success in mind—quick, tasty, realistic, yet fresh and exciting. Uncover the small appliances hiding in the cupboard that are filled with potential. Combine these into the best dinner ever placed on the fork. Finally, season it with a little love and good conversation—aspects that are easier to add when stress is minimized.

Some nights, the slow-and-steady approach wins the game. Other nights, a quick sizzle inspires the meal. Are these conflicting thoughts? No, they make up the tapestry that weaves flavors together and keeps dinner interesting, calling the family to the table. Unlock that simmering, sizzling potential! Use the timesaving wonders in these pages to create great meals. Let the slow cooker simmer or the contact grill sear. Recipes cooked this way are often easier to prepare, and taste better, than what might cook on the stove. Tips suggesting how to complete meals or streamline the work pepper these pages, as well.

You can be successful every night. Keep the mealtime monsters at bay and take charge. We feel strongly that you can have a tasty dinner on the table, often in less time than it takes to load up the car and drive to the nearby burger joint. Plus, you and your family will reap the benefits of great-tasting, nutritious meals that can be laced with memories and good conversation. The dinner table can become a refuge instead of a battleground!

Confessions of Test-Kitchen Home Economists

Dinner is a real challenge, but just because we are professionally trained home economists does not mean we win the battle every night. We are also realistic and human—and our families come first.

We first worked together in a test kitchen at Rival Manufacturing Company—maker of the Crock-Pot® slow cooker. There we tested performance, identified features, and worked with almost every small appliance one might consider, from hot dog cookers and donut bakers to slow cookers, grills, mixers, blenders, and toasters. From the superfluous to the most useful kitchen basics, we tried it all.

Twenty-five years of working, and cooking, together have taken us to food and appliance companies across the country. We have shared recipes with friendly cooks at the Iowa State Fair and food editors in New York, Chicago, and Birmingham. We have heard the laments of families at chili festivals on neighborhood streets, and talked to professionals at national appliance and food shows. The question is always the same—and that is about how to serve great-tasting, easy to prepare, new creative dishes for dinner each night.

Yet schedules, laughing (or crying) children, clubs, lessons, jobs, flights, meetings, church, and everything else that happens each day can get in the way. While we can't make all of those things go away—and honestly, we wouldn't want to, for those are the rhythms of life—we can offer a successful plan that will help you get dinner on the table. We can also share many tips about organizing, preparing, and serving food, gleamed from great test kitchens, the backs of booths at trade shows, street fairs, cooking classes, and demonstration counters, among the many other interesting venues at which we have cooked. We bring you real-life food tips that will give you insight into serving meals that are easier to prepare and taste better than you ever thought possible.

The tips that follow are our gift to you, and they work no matter what cooking method you might choose. Fill the slow cooker or heat up the grill—but make dinner preparations as fast and painless as possible. Be successful, and enjoy the experience.

A Few Tricks and Secrets We Have Learned

Organization does help. People tell us we make it look easy. Great. But "easy" is not a genetic trait. Make a list, check it twice, stock up, clean up, and plan ahead, and the meal will be easy. Just like those TV chefs who make a gourmet dinner for 1000 look like it can be done in a one-hour show, you can become your own, behind-the-scenes assistant. Think about what might be frozen, chopped, or readied ahead of time. Make a double batch. Use the tips throughout the book and you will look, and feel, like that TV star in your own kitchen.

Cooking is more art than science. Adapt, change, create. Baking is chemistry, and formulas must be exact. One the other hand, a salad is a personal preference, so go ahead and substitute one type of vinegar for another, or leave out the mushrooms. In fact, baked goods, candies, delicate sauces, and a few other rare exceptions are balanced formulas that cannot be changed, but for most other types of savory cooking, the recipe is a guide that is just waiting for you to customize it.

Even appliances have their own personalities. Some cook a little hotter, some a little cooler. Each and every slow cooker we touch is wonderful, but often each one varies just slightly from the next one on the counter. They vary by brand, model, and year produced.

Even those that look identical may cook slightly different from one another, meaning that cooking times are estimates. Choose the appliance carefully, based on features and not just cost, and then take time to bond with it and discover the traits that will lead to success.

Families vary. Each family has its own personality and quirks. Some like their food hotter, or crisper, or softer, or browner. Some find that a dish serves more people than a recipe says it will. No matter—there is no one right way. Adapt and learn to love the unique aspects and flavors your family treasures. Even we, as food professionals, have our own taste preferences. Roxanne would rarely add raisins, and Kathy would not think of rhubarb first. Kathy may choose vanilla, Roxanne chocolate. Such is the variety of life. On the other hand, don't take chances. Food safety, cleanliness, and careful preparation methods are gospels that must be followed.

Small appliances are great tools. Use small appliances for all that they offer. Count on them. Read the instructions or cookbooks that come with them. We could be prejudiced, of course, because we may have written that instruction booklet, but it really is packed with information on how to use the appliance. Always follow those guidelines.

The Yin and the Yang

The fast and the slow. The simmer and the sizzle. Conflicting? No, they are contrasts that enrich each other. Slow cookers, known for simmering rich stews and comforting soups strike a flavor balance against contact grills, which sear burgers and crisp sandwiches. Some nights, a quick, charred steak tastes perfect, while a spicy, warm chili reaches nirvana another day. Most importantly, both of these electric marvels empower you to get dinner on the table and make convenient, ready-when-you-are, flavorful meals.

Stock Up

The well-stocked pantry of today is very different from one just a few years ago, yet the theory of what it offers never changes. With basics on hand, creating a great dinner is a quicker process.

Today's list of basics is just a bit different. Early slow-cooker recipes called for salt and pepper and a lot of condensed soup. Just 30 years ago, fresh ingredients—including fresh vegetables and herbs, and the host of dried seasonings and condiments available today—were considered extravagant, gourmet cooking. Today, they are necessities for the bold, rich flavors we crave!

Pantry Tips

Find it and label it so you can use it. Keeping parts of this and that, in unmarked bags, stuffed here and there, is no way to keep a pantry, because you will never be able to find anything when you need it. Take a day to clean out, throw out, and collect similar types of ingredients into well-marked, easy-to-find places.

Check labels to review which condiments need to be refrigerated, which have expired, and which are ready for cooking.

The spice cabinet needs special attention. Herbs and spices, even dry ones in those little bottles and cans, are highly perishable. We have had the privilege of working with a seasoning company for many years, and we never cease to be amazed at the stories we hear of old, stale seasonings. One consumer recently admitted that she kept an opened herb bottle in her pantry for more than 25 years.

Herbs and spices should be fresh and aromatic; if the tantalizing smell isn't there, why even try to use it to flavor your food? Stale seasonings, especially if added to a slow cooker, will result in a bitter flavor at worst, and a tasteless dish at best. Date the bottom of the jar or can when you open it, and be skeptical if it is more than six months old. In addition, don't store your herbs and spices over the range. Pick a cool, dry spot, or refrigerate them in airtight packages.

Get out the slow cooker and contact grill. Find a convenient place to store them so they are clean and handy. Julia Child, one of the foremost mentors for cooks, kept a pegboard with all of her utensils outlined on it so she knew where each was stored. No, we don't go that far, but we do suggest that you find the utensils and tools you use daily, and store them where you can get to them easily.

An Updated Pantry

What might you find in our pantries? This is not a perfect list, but it will offer you a guide into a new world of fresh and flavorful cooking.

- Vinegars, including white wine, red wine, and balsamic
- Sun-dried tomatoes, both oil packed and not, and roasted red peppers
- Pesto and sun-dried tomato pesto
- Crushed red pepper flakes, dry minced garlic, curry powder, tarragon, and other herbs and seasonings
- Olive oil
- Rice, couscous, and pasta in various shapes
- Fresh garlic and gingerroot (for longer storage, keep ginger in the freezer)

STOCKING UP RECIPES

All-Time Favorite Pesto

2 cups fresh basil leaves

¼ cup pine nuts, toasted

1 to 2 cloves garlic, cut in half

½ teaspoon kosher salt

¼ teaspoon coarse ground pepper

⅔ cup extra virgin olive oil

½ cup freshly grated Parmesan cheese

1. Place basil, pine nuts, garlic, salt, and pepper in work bowl of food processor. Pulse to finely chop.

2. While the food processor is running, add oil in a smooth stream. Blend until smooth.

3. Place in mixing bowl and stir in cheese.

Makes about 1 cup.
Preparation time: 5 minutes

Beef Stock

TIPS

This is the time to use bone-in pieces of beef. The bones add a richer flavor to the stock.

Substitute the seasonings as you wish. Beef stock is great with garlic, or you can add basil or increase the pepper. Do not, however, overdo the salt. It is always better to under-salt the stock at this stage, so that you can salt it as you use it.

Avoid using the leaves of the celery stalk, as they will add a bitter flavor.

Makes about 7 cups.
Preparation time: 5 minutes
Cooking time: 11 to 12
  hours
Slow cooker size: Medium or
  large; round or oval

Great stock is so easy to prepare in the slow cooker. Plus, the flavor is rich and you can control the salt.

2 tablespoons vegetable oil

2 to 3 pounds meat beef bones, beef shanks, or oxtails

1 onion, cut into 8 wedges

2 carrots, cut into 2-inch pieces

2 stalks celery, cut into 2-inch pieces

6 whole black peppercorns

1 bay leaf

1 teaspoon salt

½ teaspoon leaf thyme

2 sprigs parsley

6 cups water

1. Heat oil in large, heavy skillet over medium-high heat. Add beef and cook until well browned on all sides.

2. Place beef and remaining ingredients in slow cooker. Cover and cook on low setting for 11 to 12 hours. Turn slow cooker off and allow it to cool slightly.

3. Place a fine mesh strainer over a large, deep bowl. Carefully ladle stock into strainer.

4. If desired, cover and refrigerate overnight. Skim and discard fat. Spoon into 1- or 2-cup containers; cover, label, date, and freeze.

Caramelized Onions

TIPS

Caramelized onions may be frozen. Spoon ½ to 1 cup of caramelized onions into freezer containers; label, date, and freeze.

Caramelized onions add incredible flavor to so many dishes—they can accompany a roast, top a steak, add flavor to a sandwich, finish off a pizza, or make a burger, and the list goes on!

6 medium sweet yellow onions, thinly sliced

2 tablespoons olive oil

2 tablespoons butter

Salt and pepper, to taste

1. Place onions, oil, and butter in slow cooker. Cover and cook on high setting for 6 to 8 hours.

2. Season to taste with salt and pepper.

Makes about 4 cups.
Preparation time: 5 minutes
Cooking time: 6 to 8 hours
Slow cooker size: Medium or
    large; round or oval

Chicken Stock

TIPS

This is the time to use bony pieces of chicken, as the bones add flavor.

Substitute the seasonings as you wish. Add a sprig or two of fresh parsley or tarragon, or increase the sage. Do not, however, overdo the salt. It is always better to under-salt the stock at this stage, so that you can salt it as you use it.

Avoid using the leaves of the celery stalk, as they will add a bitter flavor.

Once you taste this chicken stock, you will never want to go back to the canned versions!

2 to 3 pounds bony pieces of chicken

1 onion, cut into 8 wedges

2 carrots, cut into 2-inch pieces

2 stalks celery, cut into 2-inch pieces

4 to 5 whole black peppercorns

1 teaspoon salt

1 bay leaf

½ teaspoon dried thyme leaves

¼ teaspoon rubbed sage

6 cups water

1. Place all ingredients in slow cooker. Cover and cook on low setting for 10 to 12 hours.
2. Turn slow cooker off and allow to cool slightly.
3. Place a fine mesh strainer over a large, deep bowl. Carefully ladle stock into the strainer. If desired, cover and refrigerate overnight.
4. Skim and discard fat. Spoon into 1- or 2-cup containers; cover, label, date, and freeze.

Makes about 7 cups.
Preparation time: 5 minutes
Cooking time: 10 to
   12 hours
Slow cooker size: Medium or
   large; round or oval

Our Top Ten Quick Tips

Plan Ahead

Can't you picture it? You're driving home, talking to the baby-sitter on the cell phone, weaving in and out of traffic, knowing you still have to get the kids to practice in an hour, and your family asks, "What's for dinner?" Surprise! Dinnertime comes around about the same time each and every day. Yet, all of us have been caught off guard about what to eat. Instead, adopt a new policy. Know that your family will be hungry each night. Take a minute or two each day to think about what to fix the next day. While most of us begin to think about what to fix for dinner on the way home each night, your meal planning will be made easier and faster if you think about it a little in advance—even the night before. Then, make it a habit to always have something simmering in the slow cooker, or thawed and ready to grill, and have something fresh in the refrigerator to complete the meal before leaving for work. You will find it is so much easier to plan ahead than to create a dinner idea when you are tired and stressed. Once you have mastered the next-day planning, inch yourself along to planning a few days in advance. Soon, you will have the confidence of knowing that your week's worth of menus are planned and ready to implement.

Stay Out of the Store

While we cook all of the time, neither of us really likes shopping. Sure, sometimes we find it fascinating to wander the isles, looking at the newest innovations and soaking in the colors and smells of the fresh produce. But then reality hits, and we race on. If you shop each night, think of that 30 minutes, or longer, as wasted time. Your stomach growls, the kids whine, and you stand in line. Once you have mastered the art of planning your dinner in advance, complete the course by learning how to make a shopping list. As a bare minimum, try to avoid shopping daily; you will save both time and money. Sure, we all have to stop for one item occasionally, and we have all turned to last-minute salad bars and delis for help, but try to keep the daily shopping trips to a minimum.

Shopping Isn't Done Until the Food Is Stored Correctly

Correct storage, in this sense, doesn't just mean that food is taken out of the bag or stuffed in the refrigerator. Instead, think about what can be done ahead of time so meal preparation is efficient. Wash and dry lettuce, then store it in paper towels, in a sealed zip-top bag, so it will stay fresher, longer. In fact, wash all of the produce so it is ready to use. (The exceptions: mushrooms and berries should be washed just before using.) Separate the chicken or ground beef into usable sizes before freezing. For example, it isn't efficient or smart to thaw and maybe waste 2 pounds of ground beef when all you needed was 1 pound, just because the store had 3-pound packages on sale. Chicken breasts or pork chops can be separated into flat packages sized for each recipe before freezing, so they can be thawed and cooked quickly.

Make Wise Use of Convenience Foods

The salad bar, the olive bar, the deli's rotisserie chicken, and so many other prepared foods are mealtime treasures. They are real, time saving, honest-to-goodness foods that will add flavor and save you time. However, some boxed, bagged, or canned foods are expensive and taste like they might be better left on the shelf, while others offer "wow" flavors that enliven a quick meal. Choose wisely, then mix and match, based on what your family enjoys and what steps are worth the money for you. For example, a freshly grilled chicken breast, flavored with garlic and lemon, accompanied with a crisp salad that began as a packaged salad mix, adorned with freshly chopped red peppers from the salad bar, and served with a bakery-fresh baguette, is a dinner definitely worthy of serving.

Train the Brain

In what order do things need to be prepared? If you have made a basic plan and bought the necessary ingredients, dinner will take shape quickly. Yet, it will go even faster if you can think through the order that will make preparation easiest overall. What steps take the longest? Start the water boiling for pasta, or preheat the oven when you first walk into the kitchen. Does the slow cooker need to be turned to high, and do frozen vegetables need to be stirred into a soup or stew? Try to streamline the steps and have everything ready to serve at once.

Choose a Bottle of Flavor

Keep a selection of bottled marinades, mustards, sauces, vinegars, and such on hand to enliven even the plainest piece of meat. With meat in the center of the plate, the rest of the meal will take shape more quickly. But don't settle for plain, ho-hum meat. Add zip, zest, or a splash, and enhance the flavor. Then, look at other flavor enhancements for every course. Bottled, roasted red peppers in a salad, sun-dried tomatoes in a sauce, marinated artichokes in a casserole, and so many other enhancements can make your dishes more flavorful. Even that luscious dish simmering in the slow cooker may taste better if you stir in just a hint of that favorite sauce or a few more herbs. Quick need not taste boring.

Think Fresh for Speed

Once, opening a can may have been just the ticket to a fast dinner. Today, with the fresh produce that is available, raw foods may be faster to prepare, taste better, and offer more nutrition. Trimmed, ready-to-serve carrots and celery sticks, fresh lettuce or salad mixes, or precut vegetables and fruits are readily available and make meal preparation fast and easy. Even if you want a hot side dish to round out a meal, broccoli florets or baby spinach leaves will steam in just minutes in a covered pan on the stove.

Do It Once, Use It Twice

Always think ahead in the kitchen. Do you need a chopped onion tonight, and another one for the slow-cooker dish you will prepare tomorrow? Chop it all at once, then refrigerate what you need for tomorrow. Cook a little extra chicken—but be sure you use it up, or label and freeze it for another night. Toast some extra almonds one night, then freeze them for another day. Make extra pesto for use next week. Hard-cook eggs one night so they are chilled and ready to slice for the salad another night.

Get the Necessary Utensils and Pans

A second set of mixing bowls or another skillet just might get you moving more quickly. Don't go overboard—you don't need to stock a commercial, restaurant kitchen, and too much will just get in the way. But don't short yourself either. Have what you need so you don't have to constantly stop to wash equipment. Then, organize the equipment so it is right where you need it, ready to use.

Just Start!

The best and fastest way to a successful dinner is to just start cooking. Even small steps toward cooking will reward your family with great dinners. Plus, after each small step is completed, you will feel a bit more confident in your ability to serve a great dinner. You will move a step away from carry-out or the same old boring foods, and one step closer to quick, stress-free, tasty meals. These recipes for the slow cooker or grill are great, easy starting points. Add a crisp salad, crusty bread, or such, and dinner is on the table.

Dress Up the Dinner Table—Presentation Is Everything!

Advertisements make even the simplest item seem special, elegant, and necessary. A beautiful photo, surrounded by smiling people, with just the right lighting and music, sells a concept. In other words, presentation is everything. Does your dinner table need an advertisement? Do you feel like it needs a jazzy little jingle or an incredible picture to call the family to the table and set the mood?

When we do cooking classes or television shows, the presentation is carefully planned. Colorful dishes that contrast with the food are nestled against fabrics or props that perfectly reflect the theme. Every detail is considered when the set is chosen for a food photograph. While no one has the time, or the dishes, to set up such a display every time dinner is served, some of the same concepts can be quickly applied to your dinner table. In fact, a pretty presentation can be one of your many tricks for a quick, wonderful meal.

If you leave the food in the pan, rely on paper plates, or prop the meal on the TV tray, the meal takes on a ho-hum feel. Spruce up the appearance just a bit, and watch the family gather round and enjoy the simplest meal even more. They really will linger a bit longer. Savor the flavors and the company for just a few minutes more, and create family memories. Isn't that what mealtime should be about?

Yet, how can presentation style be improved quickly, easily, and without additional work? Try some of these professional tips.

- Purchase and use a few dramatic plates, platters, and serving bowls. They need not cost a lot. Shop the discount stores or even garage sales. Unique platters, from incomplete sets, are often sold on clearance tables at the best department stores. They don't have to match—just choose complementary colors so you can mix them up. Add a few place mats, and voilà: dinner is special.

- Think outside of the box and have fun. We have done great shows with beach towels draped across the table instead of tablecloths, and we have served picnics on denim, stir-fry on wooden mats, elegant cakes on sheer, iridescent cloths, tacos on new, colorful rugs, and tailgate parties on sheets printed with a sports theme. A yard of fabric, draped over the center of the table, adds inexpensive color.

- Find new uses for containers and bowls. Turn cake stands upside down and rest platters on top to create height and interest. Serve a main-dish salad on a large meat platter. Serve a dip or an ice cream sundae in a martini class. Stand carrots, celery, or asparagus spears upright in pretty glasses.

- Find humor at the table. Serve your very own stir-fry creation in disposable take-out boxes that resemble those from an Asian restaurant. Roxanne once hosted a birthday party for her dog, and served the people that gathered wonderful salads from a new, pretty dog dish. Serve chips out of a clean, unused cowboy boot. Serve the simplest peanut butter sandwich on your best china. We once served a tailgate party with terry-cloth hand towels instead of napkins, but we tied each one with a whistle. It got a little noisy, but the result was fun.

- Themes add to the enjoyment. If the main dish is Mexican, set a piñata on the table. You may almost feel the ocean if you arrange seashells on the table while you enjoy shrimp salad. Travel souvenirs may help you remember that perfect dish you enjoyed on vacation. Have your child arrange a collection of farm animals in the center of the table, or gather your own collection of angels. Kathy's daughter is known for hosting murder-mystery games, and the table helps her guests imagine themselves in the English countryside or the old West. Add some candles or scatter a few chocolate candies around. The very fact that your table looks distinctive will transform your everyday meal into something special.

- Get your family to help. While you prepare the dinner, children can color place mats. Teens can tie ribbons around the napkins to make napkin rings. Dad can light the candles.

Trade Secrets

Following are a few more secrets that just seem to make the dinner preparation easier and the food taste better.

- Always let roasted or grilled meat stand for 10 or 15 minutes before slicing. The juices will redistribute, and the meat will taste even more flavorful. This is true if you take it out of the slow cooker, off the grill, or any other way you cook meat.

- Freeze meat or chicken breasts for about 30 minutes before slicing or chopping. It is easier to cut ice-cold meat or chicken than it is to slice chilled meat.

- Many of the recipes in this book use a marinade to impart great flavors to the meat. Current food safety recommendations indicate that you must discard the marinade, or, if you want to use it as a brush or dipping sauce, transfer the marinade to a saucepan and heat it to a full, rolling boil.

- Fresh ginger adds a great flavor, but, if you are like most cooks, you may not use all of a fresh gingerroot before it begins to look dry or shriveled. No problem: just store it in the freezer. Slice off what you need, and return the root to the freezer. Be sure to keep it tightly sealed in plastic, and plan to use it within a year. Also, when using it, peel off just the very outer edge, as the most flavor is packed right under that skin.

- Assemble the ingredients, especially in a new recipe, before you begin cooking. The dish really will be assembled faster this way. Plus, you will avoid the scenario of having it partly mixed before realizing you are out of a critical ingredient. At the same time, just because you don't have something, don't stop—just use this as a chance to experiment and grow. Most recipes for meats, vegetables, salads, and such are guides, and they are meant for you to customize them. So go ahead and make substitutions with what you have on hand. The exception to this is breads, cakes, and baking, where exact formulas must be followed.

PART I

SLOW COOKERS

INTRODUCTION

Slow and Steady

The electric slow cooker has been a kitchen favorite since the early 1970s, when Rival Manufacturing Company electrified a bean pot and called it a Crock-Pot® slow cooker. Today, many cooks are again turning to a slow cooker for great, convenient meals, but this time, they are filled with new and bolder flavors.

What Is a Slow Cooker?

A slow cooker is a crockery or stoneware bowl, set into an electrical base, generally with heat coming evenly from all around the sides, that is designed to cook all day. A pot on the stove, a skillet turned low, or an appliance with a low setting is not a slow cooker.

What Do They Do?

- Simmer a stew, soup, chili, or Marinara Sauce
- Braise a roast
- Tenderize chops, chickens, or ribs

What They Don't Do

Contrary to what some report, they do not cook everything perfectly:

- Milk and dairy products may curdle or break down after long cooking. Add them at the very end of the cooking period.
- Rice or pasta may become too tender. Watch the cooking time.

Tips for Success

Browning and Sautéing

Once the theory was to dump in all of the ingredients and cook them all day. The complaint was that everything looked and tasted the same. Little wonder. Gone was the caramelized flavor that searing and browning produces. The recipes in this book often include steps like browning the beef and onions. Try it. The flavor is so much better. Of course, if time is limited, you can omit that step. Try great cooking techniques as the first step toward better flavor and appearance from a slow cooker.

Step Up the Herbs and Spices

Long cooking erases the distinct, pungent flavors of many herbs and seasonings. Start with double the typical amount. Taste and adjust the seasonings at the end of the cooking period. Be sure you start with a fresh jar of dried herbs, as those that are stale will become bitter after the long cooking time. Use dried leaf herbs instead of ground for the most flavor.

Timing Isn't Critical . . . Then Again, It Is

Once, we had to create every slow cooker recipe to cook for 8 to 10 hours. That was the theory, so the food would cook while people worked. Then, cooks discovered the joy and flavor of slow cooking and realized it could fit a variety of schedules. Fill and refrigerate the crock the night before, then set it in the base and turn it on as you leave your home for the afternoon. Also, slow cookers purchased today are hotter than those made in 1975. Some complain that 10 hours in their new slow cooker results in burnt sauces or dry meat. No, the exact timing of a ringing buzzer isn't there; instead, it is a relaxed and inviting cooking method. The best meals come to those who cook the food until it is done: tender and inviting, but not cooked until dry or mushy. Sometimes that may be 6 hours, sometimes 10—choose the recipe that fits your day. Remember to keep the pot covered and, if in doubt due to too crisp a carrot or rare meat, continue cooking until the meat and vegetables are tender.

Make-Ahead Convenience

The stoneware in slow cookers today is almost always removable. That means you can fill the bowl the night before, and cover and refrigerate it so it is ready to place in the electrical base on hurried mornings. The exception to this would be recipes including dry rice or pasta, as these will reconstitute before cooking. Also, potatoes (which naturally discolor after cutting) should be peeled and sliced or chopped just before cooking.

Sizes and Shapes

Today, slow cookers range from tiny, almost single-serving sizes and dip pots to larger units that hold 6 quarts or more. Those that hold between $3^1/_2$ and 5 or 6 quarts may be the best choices for most families. A round shape is great for soups and stews, while oval is especially suited to cooking roasts and ribs. Yet, most recipes will work in most shapes of slow cookers.

Some tips to remember when selecting and using your slow cooker include:

- Fill the slow cooker about half to two-thirds full.
- Always be sure the cover fits flat on the rim.
- Look for this recipe guide when selecting which slow cooker to use.
  Small round: 1–2 quarts
  Medium round or oval: 3–4 $^1/_2$ quarts
  Large round or oval: 5–6 quarts

Slow Cookers—Our Best Advice

Today, 30 years after slow cookers were first marketed, there are still some misconceptions about their use. Maybe some of our experience and suggestions will help you make the most of this convenient appliance.

- **Stirring and Peeking**

  Let the slow cooker simmer away. Do not lift the lid or stir. In fact, every time you lift the lid, a lot of valuable heat escapes. Resist the temptation to peek until close to the end of the printed cooking time. If you stir or peek, count on extending the cooking time.

- **Frozen Foods**

  Feel free to stir in frozen vegetables or shrimp at the end of the cooking time. It is best to thaw a frozen piece of meat before placing it in the slow cooker, or you must add at least $^1/_2$ cup warm liquid and extend the cooking time by several hours. As the stoneware crock is easily broken with quick temperature changes, use a little caution and plan ahead.

- **Cold Foods**

  It is fine to place very cold foods in the slow cooker, so feel free to thaw meats overnight in the refrigerator, then place them still ice cold into an unheated slow cooker and turn it on. You may find that ice-cold meat does require an extra hour or two of cooking on the low setting—just always be sure to cook it until the meat is well done and tender. In fact, this is a great tip to use if you know your slow cooker cooks a little hotter or faster, or if you want to extend the cooking time an hour or two without fear of overcooking. (Remember, if the meat is still frozen, even after being refrigerated overnight, you will need to add $\frac{1}{2}$ cup of warm water; but ice-cold meat can be placed in the slow cooker without special precautions.)

■ Cleaning

It is best to let the stoneware bowl cool a bit before filling it with hot water; then fill it with very hot, soapy water and allow it to soak a few minutes. This usually makes cleanup fairly easy. For even easier cleaning, line the slow cooker with a Reynolds® Slow Cooker Liner, a new product just introduced by Reynolds Consumer Products (a division of Alcoa Consumer Products). After cooking, just toss the plastic liner! Be sure to use only those heatproof plastic bags designed for slow cookers; do not substitute any other type of plastic food or trash bag.

■ Heating Leftovers

The microwave oven, or the conventional stovetop or oven, does a better job of heating up leftovers than the slow cooker. It is impossible to thoroughly heat the leftovers before they dry out and become overcooked in a slow cooker.

■ Thickening Juices

Since the slow cooker does not allow liquids to boil off, there will generally be more liquid at the end of the cooking time than what you started with. It will be rich, flavorful liquid.

Sometimes, however, the liquid will be even better if thickened. Try one of these methods to thicken the juices.

- Uncover the slow cooker and turn it to the high setting
  for about 30 minutes, allowing the liquids to cook down.
  (You can also transfer the juices to a saucepan on the
  stove and boil them, uncovered. This is called a "reduc-
  tion," and it intensifies the flavor of the sauce.)

- Spoon the liquids into a saucepan. Make a smooth paste of
  about 2 tablespoons of all-purpose flour and 3 to 4 tablespoons of cold water. Whisk the paste into the liquids and
  cook, stirring constantly, until liquids come to a boil and
  are thickened. (If desired, strain the liquids into the
  saucepan to remove any remaining pieces of meat, bone,
  or vegetable, then thicken.)

- Make a paste of 2 tablespoons of all-purpose flour and 3 to
  4 tablespoons of cold water, and stir into the collected
  juices in the slow cooker. Turn it to the high setting,
  cover, and allow to cook for 15 to 30 minutes. This is an
  especially good method to use for stews.

Reducing Fat

Slow cooking need not lead to a high-fat diet. Be sure to select
lean meats, then brown and drain them before placing them in
the slow cooker. To reduce fat after cooking, pour the collected juices into a large measuring cup or deep bowl and
allow them to stand a few minutes. The fat will collect at the
top, and it can then be skimmed off with a spoon.

Transporting the Slow Cooker

Slow cookers seem to multiply at potluck gatherings, tailgate
parties, and anytime one is asked to bring a dish somewhere.
They are great at keeping food hot, but sometimes, keeping
the slow cooker heated and upright seems to pose a challenge.

These tips may help. (Be sure to be very careful when moving a hot, filled slow cooker—they are heavy, and both the unit and the food will be very hot and can easily cause burns. Be sure to use hot pad holders.)

- Several companies that market slow cookers now offer carrying cases, straps to secure the covers, and other aids that might help you move the slow cooker. Check with the manufacturer to see what they offer and how to use them.

- Very large rubber bands crisscrossed over the cover and down to the handles of the slow cooker may help keep the cover secure. (This really depends on the handles of the cooker—try this method before heating the slow cooker to see if it works.)

- Find a box or plastic container just larger than the slow cooker—so the slow cooker rests flat, but there is not enough space for it to move. Line the box with heavy towels, then nestle the hot slow cooker in the towels. Fold the towels over the top of the cooker. In the car, wedge the box tightly in the trunk so the box cannot slide around.

- Remember that foods in the slow cooker will cool off if the cooker is left off. Plug in the slow cooker once you arrive at your destination. Also, be smart and remember the US Department of Agriculture (USDA) recommendation that, once the food has risen to room temperature, it should be eaten within two hours or it should be discarded. No one wants to get sick from food that has been sitting out too long.

Baking in the Slow Cooker

A slow cooker is great for baking breads and cakes, but there are some special tips that apply.

- Baking in your slow cooker will result in wonderfully moist, rich breads and cakes. Choose firmer fruit and nut cakes, chocolate cakes, pound cakes, quick breads, and yeast breads. Avoid light cakes, such as angel food or chiffon cakes.

- First, find a pan. Some companies have marketed special baking inserts for slow cookers—and if your model has one, use it and follow the company's directions. If not, an 8-x-4-inch or 9-x-5-inch loaf pan will fit into most oval slow cookers. For a large, round slow cooker, generally holding between 4 and 6 quarts, choose a 1- to 2-quart round heat-proof soufflé dish, straight-sided bowl, or casserole dish. For a cheesecake, you will want to find a 7-inch springform pan. Be sure to check the fit before beginning the recipe! Nothing is worse than having a great batter ready to bake and finding that your selected pan will not fit into the slow cooker.

- Lightly grease or grease and flour the pan, or, better yet, line the pan with parchment paper or wax paper, then grease and flour as directed. The paper can easily be peeled off of the finished cake or bread.

- To make removing the hot pan from the hot slow cooker safer and easier, take two strips of aluminum foil, each about 24 inches long, and fold them into flat "ribbons" about 2 inches wide. Place the aluminum foil strips in the slow cooker, crisscrossing them in the center of the bottom and extending the strips up the sides of the slow cooker. Place the baking pan right on top of the strips. Then, once done, the four top edges of the strip can serve as lifting handles for the pan. Still, always use hot pads and extreme caution to avoid getting burned.

- Place the filled baking pan, uncovered, directly into the slow cooker. Do not cover the baking pan or place it on a rack. The one exception to this is the cheesecake, where the filled springform pan is placed on an aluminum foil collar.

- Cover the slow cooker and bake on high setting. Do not peek until after the minimum baking time has passed.

- The recommendations for baking in slow cookers have changed over the years. Once, the baking pans designed for slow cooking had specially formed, vented covers. Otherwise, you were directed to cover the bread or cake with several paper towels as it baked. These days, as slow cookers have become hotter, this step does not seem to be required. However, if you have an older model and the top of the cake stays quite moist, you might try covering the top of the cake loosely with several paper towels. Also, baking pans could at one time be made from coffee cans or shortening cans. Now, as so many "cans" are actually made of plastic or cardboard, you will need to find a glass or metal, oven-safe pan or dish.

- Carefully, using hot pad holders, remove the baking pan and allow it to stand for 10 minutes. Turn bread or cake out of pan and allow it to cool on a wire rack.

APPETIZERS

Artichoke and Spinach Cheese Spread

This is one of our favorites, and we might argue that there is no better recipe for this dip.

TIPS

Be sure to squeeze the thawed spinach between paper towels until dry, then stir into slow cooker.

For a tasty topping, toast ½ cup of chopped pecans and sprinkle over the top of the spread just before serving.

Makes about 6 cups.
Preparation time: 10 minutes
Cooking time: 1 hour
Slow cooker size: Small or medium; round

---

2 tablespoons butter

1 medium onion, chopped

2 cloves garlic, minced

1 package (10 ounces) frozen chopped spinach, thawed and drained

1 can (13.75 ounces) artichoke hearts, drained and chopped

1 package (8 ounces) cream cheese, softened and cut into ½-in. cubes

½ cup mayonnaise

¾ cup shredded Parmesan cheese

1 package (8 ounces) shredded Colby-Monterey Jack cheese

Toasted French bread slices or tortilla chips

1. Melt butter in small skillet over medium heat. Add onion and garlic; cook, stirring frequently, until onion is tender.

2. Spoon mixture into lightly greased slow cooker. Stir in spinach and artichoke hearts.

3. Stir in cream cheese, mayonnaise, Parmesan cheese, and shredded cheese.

4. Cover and cook on low setting for 1 hour or until melted, stirring occasionally. Serve warm, as a spread on French bread slices or tortilla chips.

Beef and Cream Cheese Fondue

TIPS

Thin fondue with a little additional milk or white wine, if desired.

Recipe may be doubled for larger gatherings.

Season fondue with freshly ground pepper, if desired.

1 package (8 ounces) cream cheese, cut into cubes

½ cup milk

½ package (2.5 ounces) dried beef

2 tablespoons minced onion

1 clove garlic, minced

1 teaspoon dry mustard

Toasted baguette slices or French bread, cut into 1-inch cubes

✦✦✦

1. Spray slow cooker with nonstick spray coating. Place cream cheese and milk in slow cooker.

2. Cover and cook on low setting for 1 hour, or until cheese is melted, stirring occasionally.

3. Finely chop dried beef. Stir beef, onion, garlic, and mustard into fondue.

4. Cover and continue to cook on low setting for 30 minutes or until warm. Serve warm, with bread slices or cubes.

Makes 6 to 8 appetizer servings.
Preparation time: 5 minutes
Cooking time: 1½ hours
Slow cooker size: Small or medium; round

Christmas Eve Franks

TIPS

This recipe can be doubled to feed a larger crowd. Place in a large slow cooker and cook as directed.

This is a Wyss family tradition. We gather each Christmas Eve and create warm memories. There are always lots of appetizers for snacking, and this is an old standby!

1 package (12 ounces) cocktail wieners

½ cup ketchup

¼ cup barbecue sauce

⅓ cup bourbon

2 teaspoons Worcestershire sauce

½ cup brown sugar

✦✦✦

1. Place wieners in slow cooker.
2. Combine remaining ingredients and pour over franks; stir carefully to blend.
3. Cover and cook on high setting for 1 to 2 hours to heat through and blend flavors.
4. Turn to low setting to keep warm, then serve with toothpicks.

Makes 8 appetizer servings.
Preparation time: 5 minutes
Cooking time: 1 to 2 hours
Slow cooker size: Medium; round

Firehouse Meatballs

2½ pounds ground chuck

2 eggs

1 package (5 ounces) spicy wing mix

1 jar (12 ounces) chili sauce

1 jar (18 ounces) grape jelly

½ cup mild salsa

⅓ cup ketchup

✳ ✳ ✳

1. Combine ground beef, eggs, and spicy wing mix. Shape into 1-inch balls.

2. Place in shallow baking pans. Bake in a preheated, 400° oven for 15 minutes, or until done.

3. Drain; place in slow cooker.

4. Combine remaining ingredients and pour over meatballs. Cover and cook on high setting for 1 hour, or until sauce is hot.

5. Reduce to low setting and cook for 2 to 3 hours. Serve warm on toothpicks.

Makes approximately
   5 dozen.
Preparation time: 20 to
   25 minutes
Cooking time: 3 to 4 hours
Slow cooker size: Medium;
   round or oval

Hearts of Palm Dip

1 can (14 ounces) hearts of palm, drained and finely diced

1 cup shredded mozzarella cheese

¾ cup mayonnaise

½ cup shredded Parmesan cheese

¼ cup sour cream

2 green onions, finely minced

¼ teaspoon garlic salt

Corn chips, crackers, or pita wedges

✴✴✴

1. Combine all ingredients, except corn chips, crackers, or pita wedges.
2. Spoon into slow cooker that has been coated with nonstick spray coating.
3. Cover and cook on low setting for 1 hour.
4. Stir, then cook on low setting for 1 additional hour, or until cheeses have melted and dip is warm throughout. Serve warm with corn chips, crackers, or pita wedges.

Makes about 2 cups.
Preparation time: 10 minutes
Cooking time: 2 hours
Slow cooker size: Small;
  round

Hot Citrus-Glazed Wings

No need to go out for sweet and spicy wings! Let the slow cooker help make this an easy appetizer to share.

TIPS

If possible, purchase precut drumettes, now available at many grocery stores.

3 pounds chicken drumettes or chicken wings, cut in half with tips discarded

½ cup orange juice

½ cup honey

¼ cup soy sauce

¼ cup ketchup

6 cloves garlic, minced

2 tablespoons minced ginger

2 tablespoons hot pepper sauce

✳ ✳ ✳

1. Line a shallow baking pan with aluminum foil and spray with nonstick spray coating.
2. Arrange chicken in a single layer. Bake at 500° for about 20 minutes, or until golden brown. Drain and place in slow cooker.
3. Combine remaining ingredients and pour over chicken.
4. Cover and cook on low setting for 4 to 6 hours. Serve warm.

Makes 10 to 12 appetizer servings.
Preparation time: 20 minutes
Cooking time: 4 to 6 hours
Slow cooker size: Medium or large; round or oval

Housewares Show Favorite Crab Dip

TIPS

For an extra treat, deep-fry wonton wrappers and serve them instead of chips—the taste will remind you of Crab Rangoon. Preheat deep fryer with vegetable oil to 350°, according to manufacturer's directions. Cut each wonton wrapper in half; fry each for 30 to 60 seconds, or just until golden. Remove with slotted spoon, and drain on paper towels.

¼ cup butter, cut into pieces

1 package (8 ounces) cream cheese, softened and cut into pieces

1 can (6½ ounces) crabmeat, drained

½ teaspoon hot pepper sauce

1 teaspoon garlic powder

2 teaspoons Worcestershire sauce

Corn chips or tortilla chips

1. Spray slow cooker with nonstick spray coating. Place all ingredients, except chips, in slow cooker.
2. Cover and heat on low setting, stirring every 30 minutes, until smooth and heated through. Serve with corn chips or tortilla chips.

Makes approximately 6 to 8 appetizer servings.
Preparation time: 5 to 10 minutes
Cooking time: 1 to 2 hours
Slow cooker size: Small; round

Pizza Dip

TIPS

Instant pizza! Spoon about 1½ cups of prepared Pizza Dip onto prepared 12-inch pizza crust. Top with additional shredded cheese. Bake at 425° for about 10 to 15 minutes, or until crust is golden and cheese is melted.

This dip tastes just like a warm and wonderful pepperoni pizza. It is also great when ladled over a pizza crust and served as a traditional pizza.

1 tablespoon olive oil
¾ cup chopped onion
1 green pepper, chopped
1 cup chopped mushrooms
2 cloves garlic, minced
1 jar (26 ounces) spaghetti sauce
½ teaspoon dried oregano leaves
½ cup chopped pepperoni
½ cup shredded mozzarella cheese
Bread sticks or French bread cubes

✳✳✳

1. Heat olive oil in a skillet over medium-high heat.
2. Add onion and green pepper; cook, stirring every 2 to 3 minutes.
3. Stir in mushrooms and continue cooking, stirring frequently, until vegetables are tender. Place vegetables in slow cooker.
4. Stir in spaghetti sauce, oregano, and pepperoni. Cover and cook on low setting for 4 to 6 hours.
5. Stir in mozzarella cheese just before serving. Serve warm, with bread sticks or French bread cubes.

Makes about 3 cups.
Preparation time: 10 minutes
Cooking time: 4 to 6 hours
Slow cooker size: Small or medium; round

Slow-Cooker Party Mix

This is a recipe to serve warm at gatherings. This goes together very quickly and frees the oven for main-dish cooking. We find that guests enjoy eating it warm from the slow cooker.

TIPS

This can be made ahead and frozen until ready to serve. It is great to have on hand, ready for quick pinches. Place frozen snack mix in slow cooker and turn to low setting until mix is warm.

2 cups crispy wheat cereal squares

2 cups crispy rice cereal squares

4 cups toasted oat cereal

2 cups pretzel twists

1 cup peanuts

⅓ cup butter, melted

3 to 4 tablespoons Worcestershire sauce

½ teaspoon seasoned salt

1 teaspoon garlic powder

3 dashes hot pepper sauce

✦✦✦

1. Place cereals, pretzels, and peanuts in slow cooker. Carefully blend together.
2. Combine remaining ingredients and drizzle over cereal mixture, stirring carefully to blend.
3. Cover and cook on low setting for 2 hours, stirring every 30 minutes. Serve warm or at room temperature.

Makes approximately
3 quarts.
Preparation time: 10 minutes
Cooking time: 2 hours
Slow cooker size: Medium or
large; round or oval

Tex-Mex Spinach Dip

This is an all-time favorite. It has just four ingredients, and yet, it tastes like the dip that is served at the best Mexican restaurant in the neighborhood. We both serve it often.

TIPS

Be sure to squeeze the spinach dry before combining it with the remaining ingredients.

This tastes equally as good on fresh vegetables as it does on chips. Try it.

1 package (10 ounces) frozen chopped spinach

1 can (14.5 ounces) diced tomatoes and green chiles

1 package (8 ounces) cream cheese, softened

1 package (8 ounces) shredded Colby-Monterey Jack cheese

Tortilla chips or fresh vegetables, for dipping

1. Thaw spinach; drain well, squeezing between paper towels until dry.

2. Stir together spinach, diced tomatoes and green chiles, cream cheese, and cheese. Spoon into slow cooker that has been sprayed with nonstick spray coating.

3. Cover and cook on high setting for 30 minutes to 1 hour, or until cheese is melted, stirring occasionally.

4. Reduce to low setting for serving. Serve warm, with tortilla chips or fresh vegetables.

Makes about 4 cups.
Preparation time: 5 minutes
Cooking time: $\frac{1}{2}$ to 1 hour
Slow cooker size: Small; round

PASTA AND SAUCES

Beef and Fennel Ragu

TIPS

It is fun to use a variety of pasta shapes. The pappardelle pasta is a wide noodle with rippled sides. Mafalde is similar to a lasagna noodle, but it is narrow.

1 pound ground beef, browned and drained

1 small onion, coarsely diced

1 fennel bulb, trimmed and cut into thin strips

2 carrots, finely chopped

1 teaspoon dry minced garlic

1 jar (1 pound 10 ounces) roasted red pepper sauce

½ cup water

8 ounces pappardelle pasta or mafalde pasta

1 cup shredded mozzarella cheese

1. Place all ingredients, except pasta and cheese, into slow cooker. Cover and cook on low setting for 5 to 7 hours.
2. Cook pasta according to package directions; rinse. Place sauce on top of pasta and sprinkle with cheese.

Makes 4 servings.
Preparation time: 10 to
   15 minutes
Cooking time: 5 to 7 hours
Slow cooker size: Medium;
   round

Chicken Cacciatore

Chicken cooked in tomatoes, seasoned with herbs and capers, is just fabulous.

TIPS

If desired, add 1 to 2 cups of sliced, fresh mushrooms to the recipe. Sauté them with the onion, then transfer to slow cooker and proceed as recipe directs.

2 tablespoons olive oil

4 to 6 boneless, skinless chicken breast halves

Salt, to taste

1 medium onion, chopped

1 medium red pepper, chopped

3 carrots, chopped

1 can (28 ounces) diced tomatoes

1 can (8 ounces) tomato sauce

½ cup white wine

2 teaspoons dry minced garlic

½ teaspoon pepper

½ teaspoon crushed red pepper flakes

2 teaspoons Italian seasoning

1 bay leaf

1 to 2 tablespoons capers, drained

Hot cooked spaghetti

1 to 2 tablespoons minced fresh basil or parsley

✝✝✝

1. Heat oil in large skillet over medium-high heat. Add chicken breasts and quickly brown on each side. Season with salt and place in slow cooker.

2. Add onion to skillet drippings and cook, stirring frequently, until onion is tender. Transfer onion to slow cooker.

3. Add remaining ingredients, except spaghetti and minced fresh herbs. Cover and cook on low setting for 5 to 7 hours.

4. Remove bay leaf just before serving. Serve cacciatore over hot cooked spaghetti. Garnish with fresh minced basil or parsley.

Makes 4 to 6 servings.
Preparation time: 10 to 15 minutes
Cooking time: 5 to 7 hours
Slow cooker size: Medium or large; round or oval

Chunky Italian Sausage Pasta Sauce

Chunky and rich, this pasta sauce is packed with vegetables.

1 pound mild Italian sausage
1 medium onion, chopped
4 cloves garlic, minced
8 ounces mushrooms, sliced
1 green pepper, chopped
1 stalk celery, chopped
1 medium zucchini, chopped
1 can (28 ounces) diced tomatoes
1 can (6 ounces) tomato paste
2 teaspoons dried basil leaves
2 teaspoons dried oregano leaves
1 teaspoon dried thyme leaves
1 bay leaf
½ teaspoon salt
Freshly ground pepper, to taste
Hot cooked rigatoni or other pasta
Grated Parmesan cheese

✳✳✳

TIPS

The flavor of Italian sausage complements the vegetables. You may also use ground beef, or half ground beef and half Italian sausage, if you prefer. When using ground beef, add additional seasonings just before serving, if desired.

1. Cook sausage, onions, and garlic in large skillet over medium heat until meat is browned, stirring frequently.
2. Add mushrooms and cook for 2 to 3 minutes. Drain and place in slow cooker. Add remaining ingredients, except pasta and Parmesan cheese.
3. Cover and cook on low setting for 8 to 9 hours. Remove bay leaf just before serving. Serve over hot cooked pasta and top with grated Parmesan cheese.

Makes 6 to 8 servings.
Preparation time: 20 to 30 minutes
Cooking time: 8 to 9 hours
Slow cooker size: Medium or large; round or oval

Easy Italian-Sausage Pasta

TIPS

The flavor of the vegetables is even better when they are sautéed. However, if time is limited, omit that step. Just brown the sausage, then combine the browned sausage, vegetables, sauce, and seasonings in slow cooker; cover, and cook as directed.

1 pound Italian sausage, cut into 1- to 2-inch pieces

1 large onion, chopped

1 green pepper, chopped

8 ounces sliced mushrooms

4 cloves garlic, minced

2 jars (26 ounces each) spaghetti sauce

½ teaspoon dried basil leaves

½ teaspoon dried oregano leaves

Hot cooked pasta or grilled polenta
   (see recipe, page 349)

Shredded Parmesan cheese

1. Cook sausage in large skillet over medium heat until well browned. Drain, reserving 1 tablespoon of drippings.

2. Place sausage in slow cooker. Return reserved 1 tablespoon of drippings to skillet, and add onion, green pepper, and mushrooms. Cook, stirring frequently, until onion is tender. Stir cooked vegetables into sausage in slow cooker. Stir in spaghetti sauce, basil, and oregano.

3. Cover, and cook on low setting for 6 to 8 hours. Serve sauce over hot cooked pasta or grilled polenta. Sprinkle with Parmesan cheese just before serving.

Makes 6 servings.
Preparation time: 15 minutes
Cooking time: 6 to 8 hours
Slow cooker size: Medium or
   large; round or oval

Latin Chicken and Pasta

¼ cup olive oil

1 chicken, cut up, about 3 to 3½ pounds

Salt and pepper, to taste

1 large onion, chopped

1 jalapeno pepper, seeded

9 cloves garlic, halved

¼ cup cilantro leaves

1 medium red pepper, seeded

2 teaspoons ground cumin

½ cup pimento-stuffed Spanish olives, drained and sliced

2 tablespoons capers, drained

1 can (28 ounces) diced tomatoes

¼ cup white wine

Hot cooked spaghetti

¼ cup minced fresh cilantro

✦✦✦

1. Heat oil in large skillet over medium-high heat. Add chicken pieces, a few at a time, and cook until well browned. Season pieces generously with salt and pepper.

2. Remove chicken from skillet, leaving drippings, and place in slow cooker. Continue browning remaining pieces of chicken, adding additional oil if necessary.

3. Cook onion in drippings in skillet until onion is tender. Drain onion, then add to slow cooker.

4. Combine jalapeno pepper, garlic, cilantro leaves, and red pepper in work bowl of food processor. Pulse to coarsely chop.

5. Spoon mixture over chicken, sprinkle with cumin, then top with olives, capers, tomatoes, and wine. Cover and cook on low setting for 6 to 8 hours, or until chicken is very tender.

6. Remove chicken from sauce and allow pieces to cool slightly. Cut meat from bone; discard bone and skin. Shred chicken and return to sauce; allow to heat through. Serve chicken and sauce over spaghetti; sprinkle with fresh minced cilantro.

Makes 8 servings.
Preparation time: 20 minutes
Cooking time: 6 to 8 hours
Slow cooker size: Medium or large; round oval

Marinara Sauce

1 tablespoon olive oil

1 cup diced onion

8 cloves minced garlic

$\frac{3}{4}$ cup dry red wine

$\frac{1}{4}$ cup water

$\frac{1}{4}$ cup minced fresh parsley

1 tablespoon sugar

$1\frac{1}{2}$ teaspoons dried oregano leaves

2 teaspoons dried basil leaves

$\frac{1}{2}$ teaspoon pepper

$\frac{1}{2}$ teaspoon salt

2 bay leaves

1 can (28 ounces) crushed tomatoes

1 can (28 ounces) tomato puree

1 can (6 ounces) tomato paste

1. Heat oil in skillet over medium-high heat. Add onion and garlic, and cook until onion is tender, for about 5 minutes.

2. Add onions and remaining ingredients to slow cooker and stir to blend well.

3. Cover and cook on low setting for 7 to 9 hours, or high setting for 3 to 4 hours. Remove bay leaves just before serving.

Makes approximately 9 cups.
Preparation time: 10 minutes
Cooking time: 7 to 9 hours on low, or 3 to 4 hours on high
Slow cooker size: Medium or large; round

Meaty Mushroom Ragu

This has become a family favorite—a great way to introduce wild mushrooms to children and have them enjoy the flavors. Plus, it is a nice change from spaghetti and meatballs.

1 tablespoon olive oil

1 pound mixed mushrooms, such as button, portabella, and shitake, coarsely chopped

1 cup chopped onion

1 pound bulk Italian sausage

1 pound ground chuck

1 bottle (46 ounces) vegetable juice cocktail

1 can (6 ounces) tomato paste

1 teaspoon sugar

½ teaspoon salt

½ teaspoon coarsely ground black pepper

1 teaspoon dry minced garlic

2 teaspoons Worcestershire sauce

1 teaspoon dried basil leaves

Hot cooked extra-wide egg noodles

Sour cream

✦ ✦ ✦

1. Heat oil in large skillet over medium-high heat. Add mushrooms and onions; sauté until tender. Place in slow cooker.

2. Brown sausage and ground beef in skillet over medium-high heat; drain. Place in slow cooker.

3. Add remaining ingredients, except egg noodles and sour cream. Cover and cook on low setting for 6 to 8 hours.

4. Place hot cooked noodles in shallow bowls. Ladle mushroom ragu over noodles and dollop with sour cream.

Makes 6 servings.
Preparation time: 10 to 15 minutes
Cooking time: 6 to 8 hours
Slow cooker size: Medium; round

Sage Pasta and Chicken

2 tablespoons olive oil

1 pound boneless, skinless chicken breasts, cut into $^3/_4$-inch cubes

2 cups (8 ounces) sliced fresh mushrooms

3 cloves garlic, minced

2 tablespoons lemon juice

2 tablespoons white wine vinegar

2 tablespoons minced fresh sage

2 tablespoons minced fresh parsley

$^1/_2$ teaspoon salt

1 teaspoon coarsely ground fresh pepper

8 ounces penne, cooked according to package directions

2 tablespoons heavy whipping cream

2 tablespoons minced fresh sage

2 tablespoons minced fresh parsley

$^2/_3$ cup grated Parmesan cheese

✶ ✶ ✶

1. Heat oil in large skillet over medium-high heat. Add chicken to hot oil and cook, stirring frequently, just until chicken is white. Place chicken in slow cooker, leaving drippings.

2. Add mushrooms to drippings and cook just until tender; add to slow cooker.

3. Add garlic, lemon juice, white wine vinegar, 2 tablespoons of minced parsley, salt and pepper. Cover and cook on low setting for 3 to 4 hours.

4. Stir in cooked pasta, cream, additional sage and parsley, and Parmesan, just before serving.

Makes 6 servings.
Preparation time: 10 minutes
Cooking time: 3 to 4 hours
Slow cooker size: Medium; round

Salmon with Tomato Cream Sauce

We often travel to Chicago for an annual trade show for the appliance industry. One night we had dinner together at one of our favorite Italian restaurants, and the memory of that meal still lingers today. That luscious meal inspired this dish.

TIPS

This rich, creamy tomato sauce is also great with grilled, cubed chicken instead of the salmon. Prepare as directed, but omit the salmon. Instead, stir in fully cooked, cubed chicken, and allow to heat through.

Sun-dried tomatoes add a chewy, more intense flavor than just the canned tomatoes alone. They are available in two ways: packed in olive oil in jars, and packed dry in plastic bags. For this recipe, choose those packed in olive oil.

2 cans (28 ounces) crushed tomatoes

1 jar (8 ounces) sun-dried tomatoes in oil, chopped

$\frac{1}{3}$ cup chopped onion

2 cloves garlic, minced

1 teaspoon dried basil leaves

1 teaspoon dried oregano leaves

$\frac{1}{2}$ teaspoon salt

$\frac{1}{4}$ teaspoon crushed red pepper

$\frac{1}{2}$ cup dry white wine

1 to 1$\frac{1}{2}$ pounds of fresh boneless, skinless salmon, cut into 1-inch cubes

$\frac{1}{2}$ cup heavy whipping cream

Hot cooked penne pasta

✳✳✳

1. Place tomatoes, sun-dried tomatoes, onion, garlic, basil, oregano, salt, red pepper, and wine in slow cooker; stir well. Cover and cook on low setting for 5 to 6 hours.

2. Turn to high setting. Stir in salmon; cook for about 15 to 20 minutes, or until salmon is cooked and opaque.

3. Stir in cream. Serve over pasta.

Makes 6 to 8 servings.
Preparation time: 5 minutes
Cooking time: 5$\frac{1}{4}$ hours to
   6 hours and 20 minutes
Slow cooker size: Medium;
   round

Spaghetti with Meatballs

TIPS

If desired, cut two links of Italian sausage into slices and brown in hot oil. Add to slow cooker with meatballs.

1 pound ground beef

3 eggs, slightly beaten

1 cup fresh Italian bread crumbs

½ cup shredded Parmesan cheese

2 tablespoons minced fresh Italian parsley

1 tablespoon dried basil leaves

½ teaspoon salt

¼ teaspoon pepper

3 cloves garlic, minced

1 to 2 tablespoons olive oil

1 recipe Marinara Sauce (page 55)

1 package (16 ounces) spaghetti

✳ ✳ ✳

1. Place all ingredients, except olive oil, Marinara Sauce, and spaghetti, in a mixing bowl and blend well. Form into 2-inch (in diameter) meatballs.

2. Heat olive oil in heavy, large skillet. Add meatballs and brown well on all sides. Remove from heat; drain.

3. Prepare Marinara Sauce as directed, combining in slow cooker. (Do not precook; the meatballs and sauce cook together.)

4. Gently stir in cooked meatballs. Cover and cook on low setting for 6 to 8 hours.

5. Prepare spaghetti according to package directions; drain well. Top with marinara and meatballs.

Makes 8 servings.
Preparation time: 15 minutes
Cooking time: 6 to 8 hours
Slow cooker size: Medium or
  large; round or oval

Stroganoff Casserole

This family-style casserole captures the flavors of the classic dish, beef stroganoff.

1 boneless beef round steak, about 1½ pounds, cut into 2-x-½-inch strips

1 medium onion, chopped

2 stalks celery, chopped

1 cup sliced mushrooms

½ cup dry red wine

1 can (6 ounces) tomato paste

1 can (14.5 ounces) beef broth

1 teaspoon dry minced garlic

½ teaspoon dried oregano leaves

¼ teaspoon pepper

1 carton (8 ounces) light sour cream

¼ cup whipped cream cheese

½ cup grated Parmesan cheese

12 ounces pasta, cooked and drained

✦ ✦ ✦

1. Place beef, onion, celery, mushrooms, wine, tomato paste, broth, and seasonings in slow cooker. Stir well. Cover and cook on low setting for 6 to 8 hours.

2. Stir in sour cream and cream cheese; blend until melted and smooth. Stir in Parmesan and pasta.

Makes 6 servings.
Preparation time: 10 minutes
Cooking time: 6 to 8 hours
Slow cooker size: Medium; round

SOUPS, STEWS, AND CHILIES

Asian Style Turkey Soup

> *This soup is packed with flavor! No one will guess that this is such a healthy, low-fat soup.*

1 pound fresh turkey breast slices

½ cup diagonally sliced celery

12 large fresh mushrooms, sliced thick

2 cans (14.5 ounces each) lower-sodium, lower-fat chicken broth

1 tablespoon soy sauce

2 tablespoons freshly squeezed lemon juice

2 tablespoons cornstarch

✴✴✴

1. Cut turkey into 1-inch cubes; place in slow cooker.
2. Add remaining ingredients, except lemon juice and cornstarch. Cover and cook on low setting for 7 to 9 hours.
3. Combine lemon juice and cornstarch; blend into soup. Cover and cook on high setting for 20 to 30 minutes.

Makes 6 servings.
Preparation time: 5 to 10 minutes
Cooking time: 7 hours 20 minutes to 9½ hours
Slow cooker size: Medium; round or oval

Brisket Chili

> *Need to feed a crowd? This recipe will wow your guests and make entertaining easy.*

3 tablespoons chili powder

1 tablespoon dry minced garlic

2 teaspoons celery seed

1 teaspoon coarsely ground pepper

$\frac{1}{2}$ teaspoon seasoned salt

$\frac{1}{4}$ teaspoon cayenne pepper

1 beef brisket, about 3 to 4 pounds, trimmed

1 small onion, finely chopped

1 small green pepper, finely chopped

1 bottle (12 ounces) chili sauce

1 cup ketchup

$\frac{1}{2}$ cup barbecue sauce

$\frac{1}{3}$ cup brown sugar

$\frac{1}{4}$ cup cider vinegar

3 tablespoons Worcestershire sauce

1 teaspoon dry mustard

2 cans (15 ounces each) chili beans

✳ ✳ ✳

1. Combine chili powder, garlic, celery seed, pepper, salt, and cayenne pepper. Rub over brisket. Cut brisket into 5 or 6 pieces; place in slow cooker.

2. Combine remaining ingredients, except beans. Pour over the meat. Cover and cook on low setting for 8 to 10 hours.

3. Remove meat and cool slightly. Skim fat from juices in slow cooker. Shred meat with two forks, and return to slow cooker.

4. Stir in beans. Cover and cook on high setting for 30 minutes to 1 hour, or until heated through.

Makes 12 servings.
Preparation time: 10 minutes
Cooking time: $8\frac{1}{2}$ to 11 hours
Slow cooker size: Medium or large; round or oval

Butternut Squash Soup

1 large butternut squash
1 medium onion, chopped
½ teaspoon dried thyme leaves
½ teaspoon nutmeg
¼ teaspoon salt
¼ teaspoon pepper
2 cans (14.5 ounces each) vegetable broth
1½ cups heavy whipping cream
Dash hot pepper sauce

1. Peel squash, remove seeds, and cut into 2-inch cubes.

2. Place squash, onion, thyme, nutmeg, salt, pepper, and broth in slow cooker. Cover and cook on low setting for 7 to 8 hours, or until squash is very tender.

3. Carefully, use immersion blender to purée soup. Stir in cream and hot pepper sauce.

TIPS

If a thinner soup is preferred, thin soup with a little milk.

To cube the squash, use a large, heavy, sharp knife. It is easier to cut if heated in the microwave oven beforehand. To do so, first pierce the skin in several places, then microwave on high power (100%) for 1 to 2 minutes. Allow to stand for 3 minutes, then cut carefully.

To make Ginger Squash Soup: Add 1 tablespoon of fresh, grated ginger to recipe; omit nutmeg.

An immersion blender makes quick work of preparing a smooth soup. If you don't have an immersion blender, allow soup to cool slightly. Then ladle, in batches, into a blender and blend until smooth. When blending a hot liquid, be sure to vent the cover. Proceed as directed by blending cream and hot pepper sauce into soup.

Makes 8 to 10 servings.
Preparation time: 10 minutes
Cooking time: 7 to 8 hours
Slow cooker size: Medium or large; round

Crunchy Walnut Topping

1 package (2 ounces) English walnut pieces

1 teaspoon sugar

¼ teaspoon hot pepper sauce

1 tablespoon unsalted butter

+ + +

1. Melt butter in small skillet over medium-high heat.

2. Add walnuts, sugar, and hot pepper sauce.

3. Stirring frequently, cook until walnuts are toasted and coated, about 3 minutes. (Watch closely so they don't over-brown.)

4. Spoon onto a paper towel-lined plate to drain and cool. Sprinkle over each serving of soup.

Champion Chili

TIPS

Substitute 2 pounds of ground beef and omit sausage, if desired.

1 pound lean ground beef

1 pound bulk pork sausage

1 large onion, chopped

3 cloves garlic, minced

1 package (1 ounce) chili seasoning

1 can (15.5 ounces) chili beans (mild, medium, or hot)

1 cup picante sauce or salsa

1 can (14.5 ounces) diced tomatoes and green chiles

Shredded cheddar cheese, optional

✦✦✦

1. Cook beef, sausage, and onion in Dutch oven or large skillet, over medium-high heat, until meat is brown; drain and place in slow cooker.

2. Stir in remaining ingredients, except cheese. Cover and cook on low setting for 6 to 8 hours.

3. Ladle into bowls and garnish with cheese, if desired.

Makes 6 servings.
Preparation time: 15 minutes
Cooking time: 6 to 8 hours
Slow cooker size: Medium; round

Chicken Noodle Soup

My daughter, Grace, loves noodles and can't get enough of them. When I am making this for my family, I always use a "healthy" 3-cup portion of cooked noodles.—Roxanne

1 store-bought, fully-cooked rotisserie chicken

1 onion, diced

1 cup diced celery

1 cup diced carrot

7 cups chicken broth

1 teaspoon dried thyme leaves

½ teaspoon dried tarragon leaves

2 teaspoons dried parsley flakes

2 to 3 cups egg noodles

Salt and pepper, to taste

✳ ✳ ✳

1. Remove skin from chicken and discard. Cut meat from the bone; shred meat.

2. Place shredded chicken in slow cooker with remaining ingredients, except noodles, salt, and pepper. Cover and cook on low setting for 8 to 9 hours, or high setting for 3 to 4 hours.

3. Cook noodles according to package directions; drain. Add noodles to soup and heat through. Season to taste with salt and pepper.

Makes 6 servings.
Preparation time: 10 minutes
Cooking time: 8 to 9 hours on low, or 3 to 4 hours on high
Slow cooker size: Medium; round

Creamy Cheese Soup

Midwestern winters can be cold! This warm, creamy, satisfying soup takes only minutes to prepare before walking out the door in the morning. You and your family and friends will reap the benefits—enjoy!

2 cans (14.5 ounces each) chicken broth

1 small onion, chopped

1 large carrot, chopped

1 stalk celery, chopped

1 small red pepper, chopped

2 tablespoons butter or margarine

$\frac{1}{4}$ teaspoon salt

$\frac{1}{2}$ teaspoon coarsely ground pepper

$\frac{1}{3}$ cup all-purpose flour

$\frac{1}{3}$ cup cold water

1 package (8 ounces) cream cheese

2 cups shredded cheddar cheese

1 cup beer

Cooked, crumbled crisp bacon

Sliced green onions

✦✦✦

1. Place chicken broth, onion, carrot, celery, red pepper, butter, salt, and coarsely ground pepper in slow cooker. Cover and cook on low setting for 7 to 9 hours.

2. Combine flour and water, blending until smooth. Stir flour-water paste into soup, along with cheese and beer.

3. Cover and cook on high setting for 30 minutes, stirring once halfway through. Ladle into bowls, and top with bacon and green onions.

Makes 6 to 8 servings.
Preparation time: 10 minutes
Cooking time: $7\frac{1}{2}$ to
$9\frac{1}{2}$ hours
Slow cooker size: Medium;
round

Curried Lamb Stew

TIPS

This stew is great served in bowls, with a crusty bread. It is also great served over hot cooked rice or couscous.

Leave the sweet potatoes in larger chunks so they don't overcook. If you want to extend the cooking time even longer, add sweet potatoes, cut into 1-inch cubes, during the last 2 to 3 hours.

¼ cup all-purpose flour

½ teaspoon salt

¼ teaspoon pepper

2 pounds boneless lamb, cut into 1-inch cubes

2 tablespoons vegetable oil

1 large onion, chopped

¼ cup dry white wine

3 medium carrots, sliced

2 medium sweet potatoes, peeled and cut into 2- to 3-inch pieces

1 teaspoon ground cumin

1 teaspoon curry powder

½ teaspoon turmeric

½ teaspoon dry minced garlic

½ teaspoon salt

¼ teaspoon pepper

1 can (14.5 ounces) chicken broth

1 can (15 ounces) garbanzo beans, rinsed and drained

✳✳✳

1. Combine flour, ½ teaspoon of salt, and ¼ teaspoon of pepper in zip-top bag. Add lamb cubes; seal, and shake to coat well.

2. Heat oil in large, heavy skillet over medium-high heat. Add lamb, and cook until browned. Remove lamb from skillet using a slotted spoon, leaving drippings, and place lamb in slow cooker.

Makes 6 servings.
Preparation time: 20 minutes
Cooking time: 7 to 8 hours
Slow cooker size: Medium or large; round or oval

(continues on next page)

3. Add onion to skillet and cook, stirring frequently, until onion is tender. Add wine to skillet and stir up any browned particles. Spoon onion-wine mixture into slow cooker.

4. Add remaining ingredients, except garbanzo beans. Stir well. Cover and cook on low setting for 7 to 8 hours.

5. Stir in garbanzo beans and allow to heat through for 5 to 10 minutes.

French Onion Soup

> *Caramelizing the onions in a slow cooker makes this dish so easy to prepare, you can serve it often.*

TIPS

Cover the baking sheets with aluminum foil before arranging bread on them; this will make cleanup easy, since cheese is sure to drip onto the baking sheet.

6 medium sweet yellow onions, thinly sliced

3 tablespoons olive oil

2 tablespoons butter

$\frac{1}{4}$ teaspoon dried thyme leaves

3 cans (14. 5 ounces each) beef broth or $5\frac{1}{4}$ cups of beef stock, at room temperature

2 tablespoons dry sherry

Salt and freshly ground pepper, to taste

6 to 8 slices French bread, about $\frac{3}{4}$ inch thick

6 to 8 slices Gruyère or Swiss cheese

✦✦✦

1. Place onions, oil, butter, and thyme in slow cooker. Cover and cook on high setting for 6 to 8 hours, or until onions are golden brown in color.

2. Add broth, sherry, salt, and pepper; cover and cook on high for 30 minutes.

3. Meanwhile, arrange French bread on aluminum-foil-covered baking sheet. Toast in broiler until golden brown on one side. Turn bread over, top with cheese, and broil until cheese is melted and bubbly.

4. To serve, ladle soup into serving bowls. Top each with a slice of bread covered with melted cheese.

Makes 6 to 8 servings.
Preparation time: 5 minutes
Cooking time: $6\frac{1}{2}$ to $8\frac{1}{2}$ hours
Slow cooker size: Medium or large; round or oval

Herbed Bistro Beef on Herb Cheese Biscuits

2 pounds beef stew meat

¼ cup all-purpose flour

1 teaspoon pepper

½ teaspoon salt

2 tablespoons vegetable oil

1 large onion, chopped

4 cloves garlic, minced

2 teaspoon dried rosemary leaves

2 teaspoon dried oregano leaves

2 teaspoons dried basil leaves

2 bay leaves

1 can (14.5 ounces) diced tomatoes

1 can (6 ounces) tomato paste

1 can (10 ounces) beef consommé

3 medium carrots, sliced

8 ounces mushrooms, sliced

½ teaspoon pepper

½ teaspoon salt

2 cups frozen peas, corn, or cut green beans, or a combination

Herb Cheese Biscuits (recipe following)

✦✦✦

Makes 6 to 8 servings.
Preparation time: 20 minutes
Cooking time: 8 to 10 hours
Slow cooker size: Medium or large; round or oval

1. Place meat in zip-top bag. Add flour, 1 teaspoon of pepper, and ½ teaspoon of salt; shake to coat well.

2. Heat oil in large skillet over medium-high heat. Add beef, and cook until meat is browned. Drain and transfer to slow cooker.

3. Add remaining ingredients, except frozen vegetables and biscuits. Cover, and cook on low for 8 to 10 hours.

4. About 30 minutes before serving, stir in frozen vegetables. Remove bay leaf before serving. Serve over split, warm Herb Cheese Biscuits.

Herb Cheese Biscuits

1 1/2 cups baking mix

1/2 cup milk

1/4 cup grated cheddar cheese

1/4 teaspoon dried basil leaves

1/4 teaspoon dried oregano leaves

3 tablespoons melted butter

1/4 teaspoon garlic powder

✦ ✦ ✦

1. Preheat oven to 450°. Combine baking mix, milk, cheese, basil, and oregano. Stir just until blended and a soft dough forms.
2. Drop by heaping tablespoonfuls onto greased baking sheet. Bake for 8 to 10 minutes, or until golden.
3. Blend together melted butter and garlic powder; brush over warm biscuits. Serve biscuits warm.

Makes 6 to 8 biscuits.

Hot and Spicy Chili Casserole

While this recipe has a slightly unique approach, many folks enjoy topping chili with chopped onions. This time, substitute canned French-fried onions for a crunchy treat.

8 ounces bacon

2 pounds lean ground beef

2 medium onions, chopped

½ cup chopped green or red pepper

1 can (6 ounces) tomato paste

1 can (10 ounces) mild enchilada sauce

1 can (15.5 ounces) chili beans (mild, medium, or hot)

½ cup chopped pickled jalapeno peppers, drained

1 tablespoon brown sugar

1 package (1 ounce) chili seasoning

1 can (2.8 ounces) French-fried onions

8 ounces process cheese, cubed

1. Fry bacon in a Dutch oven over medium heat until bacon is crisp. Remove bacon and reserve; measure and reserve ¼ cup of bacon drippings.

2. Pour reserved drippings back into Dutch oven and add ground beef, onions, and chopped green or red peppers. Cook over medium heat for about 5 minutes, or until meat is browned. Drain and place in slow cooker.

3. Stir in tomato paste, enchilada sauce, beans, jalapeno peppers, brown sugar, and chili seasoning. Cover and cook on low setting for 6 to 8 hours.

4. Sprinkle French-fried onions and cheese cubes over top of chili. Crumble bacon and sprinkle over top of cheese. Cover and cook for 10 to 15 minutes, or just until cheese is melted.

Makes 6 to 8 servings.
Preparation time: 15 minutes
Cooking time: 6 to 8 hours
Slow cooker size: Medium; round

Mushroom Barley Soup

The flavor of this soup is rich, satisfying, and surprising. The combination of the fresh and dried mushrooms, parsnips, and dill is worth a try.

2 tablespoons butter or margarine

1 large onion, chopped

3 cloves garlic, minced

1 pound fresh mushrooms, sliced

1 package (3.5 ounces) dried porcini mushrooms (about $\frac{1}{2}$ cup of dry mushrooms)

2 medium carrots, chopped

1 stalk celery, chopped

2 parsnips, chopped

$\frac{1}{2}$ cup pearl barley

2 cans (14 ounces each) vegetable broth

2 cups water

1 teaspoon salt

1 teaspoon dried dill weed

$\frac{1}{2}$ teaspoon pepper

$\frac{1}{4}$ cup minced fresh dill

✦✦✦

1. Melt butter in large skillet over medium-high heat. Add onion and garlic, and cook, stirring frequently, until onion is just tender.

2. Add fresh, sliced mushrooms and cook until mushrooms are tender; spoon into slow cooker.

3. Stir in remaining ingredients, except fresh dill. Cover and cook on low setting for 7 to 9 hours. Stir in fresh dill just before serving.

Makes 8 to 10 servings.
Preparation time: 10 minutes
Cooking time: 7 to 9 hours
Slow cooker size: Medium or large; round or oval

Pasta Fagioli Soup

1½ pounds ground beef

3 cloves garlic, minced

1 medium onion, chopped

2 medium carrots, chopped

2 cans (14.5 ounces each) diced tomatoes

2 cans (14.5 ounces each) beef broth

2 cups water

¼ teaspoon salt

¼ teaspoon pepper

2 teaspoons dried basil leaves

1 teaspoon dried oregano leaves

4 to 5 drops hot pepper sauce

1 small zucchini, chopped

1 can (15 ounces) red kidney beans, rinsed and drained

1 can (15 ounces) cannellini beans or Great Northern beans, rinsed and drained

2 tablespoons minced fresh basil

¾ cup small shell macaroni

Grated Parmesan cheese

✶ ✶ ✶

1. Cook ground beef in skillet over medium-high heat until browned, stirring to crumble; drain.

2. Place cooked beef, garlic, onion, carrots, tomatoes, broth, water, salt, pepper, basil, oregano, and hot pepper sauce in slow cooker. Cover and cook on low setting for 7 to 8 hours.

3. Stir in zucchini and beans, and cook on low for 1 hour.

4. Turn slow cooker to high. Stir in fresh basil and macaroni. Cover and cook on high setting for 30 minutes, or until pasta is cooked.

5. Season to taste with additional salt, pepper, and hot pepper sauce, if desired. Ladle into bowls and top each with freshly grated Parmesan cheese.

Makes 10 servings.
Preparation time: 10 to 15 minutes
Cooking time: 8½ to 9½ hours
Slow cooker size: Large; round or oval

Pork Stew with Cilantro

2 tablespoons all-purpose flour

1 teaspoon salt

$\frac{1}{4}$ teaspoon pepper

2 pounds boneless pork loin roast, cut into 1-inch cubes

2 tablespoons vegetable oil

1 large onion, chopped

3 medium carrots, sliced

8 new red potatoes, halved

1 tablespoon dry minced garlic

3 tablespoons tomato paste

1 teaspoon paprika

$\frac{1}{2}$ teaspoon cumin

$\frac{1}{2}$ teaspoon ground pepper

$\frac{1}{4}$ teaspoon cayenne pepper

1 can (14.5 ounces) beef broth

2 tablespoons minced cilantro

1. Combine flour, salt, and pepper in zip-top bag. Add pork cubes; seal and shake to coat evenly.

2. Heat oil in large skillet over medium-high heat. Add pork and cook until browned; drain and place in slow cooker.

3. Add remaining ingredients, except cilantro. Cover and cook on low setting for 6 to 7 hours. Stir in cilantro and cook for 30 minutes.

Makes 6 to 8 servings.
Preparation time: 10 to
15 minutes
Cooking time: $6\frac{1}{2}$ to
$7\frac{1}{2}$ hours
Slow cooker size: Medium or
large; round or oval

Ranch House Smokin' Chili

1 ½ pounds ground beef

1 pound spicy smoked sausage links, chopped into bite-size pieces

2 cloves garlic, minced

1 cup chopped onion

2 packages (1 ounce each) chili seasoning

½ teaspoon fennel seeds, crushed

1 bay leaf

½ cup smoky barbecue sauce

2 cans (15.5 ounces each) chili beans (mild, medium, or hot)

1 cup chunky salsa

2 cans (14.5 ounces each) diced tomatoes

2 teaspoons brown sugar

Salt and pepper, to taste

1. Cook ground beef, sausage, garlic, and onion in Dutch oven over medium-high heat, stirring frequently, until beef is browned; drain and place in slow cooker.

2. Stir in remaining ingredients. Cover and cook on low setting for 6 to 8 hours. Remove bay leaf before serving.

Makes 6 to 8 servings.
Preparation time: 10 to 15 minutes
Cooking time: 6 to 8 hours
Slow cooker size: Medium; round

Red, White, and Blue Chili

Red and white beans in the chili and a topping of crunchy blue corn chips make this a unique, great chili.

1 tablespoon vegetable oil

1 cup chopped celery

2 cups chopped green pepper

2 cups chopped onion

2 cans (14.5 ounces each) chicken broth

2 cups water

1 pound fully cooked, smoked sausage, sliced ½ inch thick

1 pound fully cooked ham, cut into ½-inch cubes

1 package (1 ounce) chili seasoning

½ teaspoon cayenne pepper

1 can (15 ounces) red kidney beans, drained

1 can (15 ounces) Great Northern beans, drained

1 bag (8 ounces) blue corn tortilla chips

+ + +

1. Heat oil in 5-quart Dutch oven. Sauté celery, green pepper, and onion until tender. Spoon vegetables into slow cooker.

2. Add remaining ingredients, except tortilla chips. Cover and cook on low setting for 6 to 8 hours.

3. Ladle into bowls and top with blue corn tortilla chips.

Makes 10 to 12 servings.
Preparation time: 10 minutes
Cooking time: 6 to 8 hours
Slow cooker size: Medium; round

Roasted Corn and Red Pepper Soup

1 package (16 ounces) frozen corn, partially thawed

1 medium onion, chopped

1 carrot, diced

3 cloves garlic, minced

1 cup chopped roasted red pepper, drained

2 cans (14.5 ounces each) diced tomatoes

½ cup white wine

1 can (14.5 ounces) chicken broth

2 cups water

½ teaspoon dried oregano leaves

½ teaspoon dried basil leaves

¼ teaspoon salt

¼ teaspoon pepper

¼ teaspoon hot pepper sauce

2 tablespoons minced fresh cilantro

✦✦✦

1. Preheat oven to 425°. Spread corn in a single layer on a 10-x-15-inch baking sheet. Bake for 20 minutes, or until corn is golden brown.

2. Place all ingredients, except cilantro, in slow cooker. Stir in roasted corn.

3. Cover and cook on low setting for 8 to 9 hours. Stir in cilantro just before serving.

Makes 8 to 10 servings.
Preparation time: 25 minutes
Cooking time: 8 to 9 hours
Slow cooker size: Medium or
  large; round or oval

Rustic Italian Beef Ragu

TIPS

It is quick and easy to toast thickly sliced bread on a contact grill. Lightly butter bread, if desired, then grill for just 1 to 2 minutes, or until toasted.

Makes 10 to 12 servings.
Preparation time: 30 to
    45 minutes
Cooking time: 7½ to
    8½ hours
Slow cooker size: Medium or
    large; round or oval

3 slices bacon

3 pounds beef stew meat

1 medium onion, chopped

8 ounces sliced fresh mushrooms

1 can (14.5 ounces) diced tomatoes

2 bay leaves

1 can (10 ounces) beef consommé

½ cup red wine

1 tablespoon dry minced garlic

1 tablespoon balsamic vinegar

½ teaspoon dried thyme leaves

¼ teaspoon salt

½ teaspoon coarsely ground black pepper

2 tablespoons all-purpose flour

¼ cup cold water

10 to 12 slices Italian or sourdough bread, about ¾ inch thick, toasted

✦✦✦

1. Fry bacon in large skillet over medium-high heat, until bacon is crisp. Remove bacon to paper-towel-lined plate; reserve drippings in skillet.

2. Add beef in batches to hot fat, and quickly brown well. Remove browned beef with slotted spoon, leaving drippings, and place in slow cooker.

3. Brown remaining beef and place in slow cooker. Add chopped onions and mushrooms to drippings in skillet and cook, stirring, until onion is translucent. Place onions and mushrooms in slow cooker.

(continues on next page)

4. Crumble bacon and add to beef. Add diced tomatoes, bay leaves, consommé, red wine, garlic, balsamic vinegar, thyme, salt, and pepper. Cover and cook on low setting for 7 to 8 hours.

5. Turn cooker to high. Remove and discard bay leaves. Blend together flour and $\frac{1}{4}$ cup of cold water, stirring to form a smooth paste. Stir flour paste into stew.

6. Cook for about 30 minutes, or until bubbly and slightly thickened. Spoon ragu over toasted bread.

Shrimp Creole

TIPS

Frozen shrimp thaws quickly; follow directions on package.

¼ cup vegetable oil

¼ cup all-purpose flour

1 large onion, chopped

2 stalks celery, chopped

1 green pepper, chopped

¼ cup red wine

5 cloves garlic, minced

1 can (14.5 ounces) diced tomatoes

1 can (8 ounces) tomato sauce

1 can (6 ounces) tomato paste

1 tablespoon Worcestershire sauce

1 teaspoon salt

½ teaspoon pepper

½ teaspoon cayenne pepper

¼ teaspoon hot pepper sauce

2 packages (12 ounces each) frozen, peeled, uncooked shrimp, thawed (or 2 pounds of fresh shrimp, peeled and deveined)

Hot cooked rice

✦✦✦

1. Heat oil in large, heavy skillet over medium-high heat. Add flour and cook, stirring, for about 5 minutes or until mixture becomes a deep, brown color.

2. Add onion, celery, and green pepper, and cook for 3 to 4 minutes, stirring frequently. Turn off heat.

3. Stir in wine, blending well. Spoon into slow cooker. Stir in garlic, tomatoes, tomato sauce, tomato paste, Worcestershire sauce, and seasonings. Cover and cook on low setting for 6 to 8 hours.

4. About 30 minutes before serving, turn slow cooker to high setting. Stir in shrimp. Cover and cook on high for about 30 minutes, or until shrimp turn pink. Serve over rice.

Makes 6 servings.
Preparation time: 20 minutes
Cooking time: 6½ to
    8½ hours
Slow cooker size: Medium or
    large; round

Steak Fajita Chili

TIPS

Brown the meat in two or three batches so as to not crowd the skillet. Crowding the meat will cause it to steam instead of browning nicely.

It is easier to thinly slice ice-cold meat. Place the meat in the freezer for about 30 minutes, or just until the meat is ice cold, then slice thinly.

> *Two great favorites, chili and fajitas, in one dish! Serve this in bowls, or spoon it into a warm tortilla. Either way, it is a winner.*

1 tablespoon vegetable oil

2 pounds beef flank or skirt steak, cut into thin strips

1 medium onion, finely chopped

1 green pepper, cut into strips

1 can (14.5 ounces) diced tomatoes and green chiles

1 can (15.5 ounces) black beans, rinsed and drained

1 package (1 ounce) chili seasoning

½ teaspoon salt

½ cup sour cream

½ cup guacamole

½ cup salsa or pico de gallo

Tortilla chips or warm, flour tortillas, optional

1. Heat oil in large skillet over medium heat. Cook steak strips until browned, stirring frequently to brown evenly; drain.

2. Place meat in slow cooker. Stir in onion, green pepper, diced tomatoes and green chiles, black beans, chili seasoning, and salt. Cover and cook on low setting for 7 to 9 hours.

3. Serve, topping each bowl with a dollop of sour cream, guacamole, and salsa. Serve with tortilla chips or spoon into a warm, flour tortilla, if desired.

Makes 8 servings.
Preparation time: 20 minutes
Cooking time: 7 to 9 hours
Slow cooker size: Medium; round

Steak Soup

TIPS

One key to this great soup is the beef base. Look for it in the grocery store with the spices and seasonings, or with the bouillon. Bouillon cubes just don't offer the rich taste or dark color of beef base paste. It does contain salt, so it is best to taste the soup after cooking and add additional salt then, if desired.

Kansas City is known for great steaks, but the name of this soup is misleading. Steak soup is served at many restaurants in Kansas City, but it is rarely made from chopped steak. Instead, it is generally made from ground beef. No matter the meat choice, it is a thick, rich soup that will warm you on a cold night.

2 pounds ground beef

1 large onion, chopped

2 stalks celery, chopped

2 large carrots, chopped

1 medium potato, peeled and chopped

1 can (14.5 ounces) diced tomatoes

1 can (14.5 ounces) beef broth

2 to 4 tablespoons beef base paste

$\frac{1}{4}$ teaspoon salt

$\frac{1}{2}$ teaspoon coarsely ground pepper

2 cups water

2 cups frozen mixed vegetables

3 tablespoons all-purpose flour

$\frac{1}{4}$ cup cold water

1. Cook ground beef in skillet over medium-high heat until browned, stirring to crumble. Drain, and place in slow cooker.

2. Stir in onion, celery, carrots, potato, tomatoes, broth, beef base, salt, and pepper. Stir in water. Cover, and cook on low setting for 7 to 9 hours.

Makes 6 to 8 servings.
Preparation time: 15 minutes
Cooking time: 8 to 10 hours
Slow cooker size: Medium or
    large; round or oval

(continues on next page)

3. Turn slow cooker to high setting. Stir in frozen vegetables, and cook for 30 minutes.

4. Stir together flour and water, making a smooth paste, then stir into soup. Continue cooking for about 30 minutes, or until thick. Taste, and add additional salt and pepper, if desired.

Texas-Style Soup

TIPS

To fry corn tortilla strips, slice corn tortillas into ½-inch strips. Pour vegetable oil into a large skillet, to a depth of about ½ inch; heat over medium-high heat until oil is hot. Add tortilla strips and cook for about 1 minute, or until crisp, turning to brown evenly. Drain on paper-towel-lined plate.

Soups are great to freeze—then just quickly heat and serve on a really busy night. Freeze the soup before topping with garnishes, croutons, or, in this case, sour cream, tortilla chips, and cheese. For even more convenience, freeze the soup in individual-size portions. For example, freeze 1 to 2 cups of soup in small, resealable bowls. Be sure

This is fast to put together, and every time I prepare it, I hear, "Wow! What is that smell? What are we having for dinner?" as soon as Grace walks in the door from school. It is a big hit!

1 pound ground chuck, browned and drained

2 cans (15 ounces each) black beans, rinsed and drained

1 can (14.5 ounces) diced tomatoes

1 can (14.5 ounces) diced tomatoes and green chiles

1 can (14.5 ounces) chicken broth

1½ cups frozen corn

2 cans (4 ounces each) canned whole green chiles, drained and coarsely chopped

4 greens onions, sliced

3 tablespoons chili powder

2 teaspoons cumin

1 teaspoon dry minced garlic

Sour cream

Tortillas or home-fried corn tortilla strips

Shredded cheddar cheese

✦ ✦ ✦

(continues on next page)

to label each container and cover tightly. If possible, place it in the refrigerator the night before you plan to use it. But, in a pinch, you can quickly microwave the frozen container. Even if you need more than one serving, microwaving a couple smaller containers is faster than heating one large one. Plus, frozen soup containers are great to take to the office for a hot lunch.

1. Place all ingredients, except sour cream, chips, and cheese, in slow cooker.

2. Cover and cook on low setting for 7 to 9 hours. Ladle into bowls and dollop with sour cream. Crunch tortilla chips (or place corn strips) on top, and sprinkle about 1 tablespoon of shredded cheese over each serving.

Makes 6 servings.
Preparation time: 10 minutes
Cooking time: 7 to 9 hours
Slow cooker size: Medium or
    large; round or oval

Two-Onion Beef Stew

2 tablespoons butter or margarine

1 package (8 ounces) sliced mushrooms

1 medium onion, coarsely diced

4 bacon slices, diced

$\frac{1}{2}$ cup all-purpose flour

$\frac{1}{2}$ teaspoon salt

$\frac{1}{2}$ teaspoon coarsely ground pepper

$1\frac{1}{2}$ pounds beef stew meat

1 cup frozen pearl onions

6 carrots, sliced

2 medium potatoes, diced

1 can (14.5 ounces) beef broth

1 cup dry red wine

1 teaspoon dry minced garlic

$\frac{1}{2}$ teaspoon dried thyme leaves

✦✦✦

1. Melt butter in large skillet; add mushrooms and coarsely diced onion. Cook until tender; remove and place in slow cooker.

2. Add bacon to skillet and brown until very crisp; remove from skillet and add to slow cooker.

3. Place flour, salt, and pepper in zip-top bag. Add beef, and shake to coat.

4. Brown meat in bacon drippings; add to slow cooker along with remaining ingredients. Cover and cook on low setting for 8 to 10 hours.

Makes 4 to 6 servings.
Preparation time: 15 to 20 minutes
Cooking time: 8 to 10 hours
Slow cooker size: Medium or large; round or oval

Vegetable Bean Soup with Parmesan Garlic Croutons

This is a great vegetarian soup—it is packed with so much flavor, no one will miss the meat.

1 medium onion, chopped

1 stalk celery, chopped

2 medium carrots, chopped

1 medium potato, peeled and chopped

4 cloves garlic, minced

2 cans (14 ounces each) vegetable broth

1 can (14.5 ounces) diced tomatoes

1 can (15 ounces) dark red kidney beans, rinsed and drained

1 can (15 ounces) black beans, rinsed and drained

1 teaspoon Italian seasoning

$\frac{1}{4}$ teaspoon crushed red pepper flakes

$\frac{1}{2}$ teaspoon salt

$\frac{1}{4}$ teaspoon pepper

1 cup coarsely chopped zucchini

2 tablespoons minced fresh parsley

1 cup frozen Italian green and yellow beans, or cut green beans

Parmesan Garlic Croutons (recipe following)

✳✳✳

1. Place all ingredients, except zucchini, parsley, frozen beans, and croutons, in slow cooker; stir well.

2. Cover, and cook on low setting for 8 to 10 hours. Stir in zucchini, parsley, and frozen beans during last 1 to 2 hours of cooking.

3. Ladle into bowls, and top each serving with Parmesan Garlic Croutons.

Makes 6 to 8 servings.
Preparation time: 10 to 15 minutes
Cooking time: 8 to 10 hours
Slow cooker size: Medium or large; round or oval

Parmesan Garlic Croutons

3 cups French bread cubes, cut into about ¾-inch cubes

3 tablespoons melted butter

1 teaspoon garlic salt

1 teaspoon dried basil leaves

3 tablespoons grated Parmesan cheese

✦ ✦ ✦

1. Preheat oven to 425°. Cover a baking sheet with aluminum foil. Place bread cubes in zip-top bag. Drizzle with melted butter, and sprinkle with garlic salt and basil; seal, and shake bag to coat bread cubes evenly.

2. Spread bread cubes on baking sheet. Bake for 10 to 15 minutes, or until bread cubes are golden and crisp. Immediately sprinkle hot bread cubes with Parmesan cheese; allow to stand for 2 to 3 minutes before serving.

Welcome Home Chili

This great-tasting chili is packed with beans and flavor. In fact, no one will suspect that you are able to stretch just 2 pounds of beef to serve so many people.

2 pounds ground beef

1 large onion, chopped

1 medium green pepper, chopped

2 cans (14.5 ounces each) diced tomatoes and green chiles

2 cans (15.5 ounces each) chili beans (mild, medium, or hot)

2 packages (1 ounce each) chili seasoning

1 can (8 ounces) tomato sauce

1 can (15 ounces) pork and beans

1 can (16 ounces) refried beans

1 can (15.5 ounces) black beans, rinsed and drained

1 can (15 ounces) dark red kidney beans, rinsed and drained

¼ cup brown sugar

¼ cup ketchup

1 teaspoon salt

✛ ✛ ✛

1. Cook beef, onion, and green pepper in Dutch oven over medium heat, stirring frequently, until beef is browned and vegetables are tender; drain and place in slow cooker.

2. Stir in remaining ingredients. Cover and cook on low setting for 6 to 8 hours.

Makes 14 to 16 servings.
Preparation time: 10 to 15 minutes
Cooking time: 6 to 8 hours
Slow cooker size: Large; round or oval

BEEF

Barbecued Brisket

TIPS

For even more Kansas City style flavor, refrigerate the cooked meat and collected juices for several hours or until well chilled. Thinly slice the meat, then reheat meat and juices until warm. Serve warm, with bread, and drizzle with additional barbecue sauce. Accompany the meal with creamy coleslaw and baked beans.

This brisket is definitely "Kansas City style." No, it is not smoked, as the great barbecue houses in Kansas City do, but the flavor of the sauce shouts of the sweet, spicy sauce found only in Kansas City. For us, two Kansas City natives, there simply isn't another kind of barbecue flavor.

1 teaspoon paprika
1 teaspoon salt
½ teaspoon pepper
1 tablespoon dry minced garlic
1 beef brisket, about 3 pounds, well trimmed
2 tablespoons vegetable oil
⅓ cup ketchup
¼ cup brown sugar
2 tablespoons molasses
1 tablespoon Worcestershire sauce

✦ ✦ ✦

1. Combine paprika, salt, pepper, and garlic; rub into both sides of brisket.
2. Heat oil in large skillet over medium-high heat. Add brisket and sear on both sides. Place brisket in slow cooker.
3. Combine ketchup, brown sugar, molasses, and Worcestershire sauce; drizzle over meat. Cover and cook on low for 8 to 10 hours.

Makes 8 servings.
Preparation time: 15 to
   20 minutes
Cooking time: 8 to 10 hours
Slow cooker size: Medium or
   large; round or oval

Bourbon-Barbecued Beef Sandwiches

Shredded beef, piled on toasted buns, is one of the all-time favorite dinners. This version is especially wonderful to take to a tailgate party or to serve while gathered in front of the television to watch a big game.

TIPS

If a chuck or arm roast is cooked until quite tender, it will easily shred using the tips of two forks. This cut is generally inexpensive and readily available, making it the perfect choice for this recipe.

2 tablespoons vegetable oil

1 boneless beef chuck or arm roast, about 3 pounds

1 medium onion, sliced

½ teaspoon salt

½ teaspoon pepper

1 tablespoon dry minced garlic

¼ cup bourbon

1 can (10½ ounces) beef consommé

1 cup barbecue sauce

Hamburger buns, split and toasted

✳ ✳ ✳

1. Heat oil in large skillet over medium-high heat. Add roast and cook until well browned on each side.

2. Place onion in slow cooker. Place meat on top of the onion and sprinkle with salt, pepper, and garlic. Pour bourbon and consommé over meat.

3. Cover and cook on low setting for 9 to 10 hours, or until beef is very tender.

4. Lift beef from slow cooker; drain and reserve liquid. Trim away fat; then, using two forks, shred the beef. Return beef to slow cooker.

5. Skim fat from collected liquid; measure out ¼ cup of collected liquid and pour over beef. Stir in barbecue sauce.

6. Cover and cook on low setting for 30 minutes to 1 hour, or until heated through. Serve on buns.

Makes 8 servings.
Preparation time: 15 minutes
Cooking time: 9½ to
   11 hours
Slow cooker size: Medium or
   large; round or oval

Chipotle Pepper Beef with Fresh Tomato Salsa

Fresh cilantro, garlic, and the unmistaken flavor of a chipotle pepper season this favorite beef cut as it simmers all day. The fresh tomato salsa provides a flavor and nutritional punch. It's a meal in itself!

TIPS

Chipotle peppers are smoked jalapeno peppers, and are often found canned in adobe sauce. Use one for this recipe, then package the rest and freeze it for use another time. For even more convenience, freeze the remaining canned chipotle peppers in very small containers so you can just thaw and use the amount you need.

Makes 6 to 8 servings.
Preparation time: 15 to
   20 minutes
Cooking time: 8 to 9 hours
Slow cooker size: Medium;
   round

1 tablespoon vegetable oil
1 beef brisket, about 2 pounds, well trimmed
1 onion, thinly sliced
6 cloves garlic, minced
1 canned chipotle pepper in adobe sauce, chopped
¼ cup minced fresh cilantro
½ teaspoon seasoned salt
½ cup tomato juice
Fresh Tomato Salsa (recipe following)
Flour tortillas, warmed

1. Cut meat into fourths. Heat oil in large skillet over medium-high heat. Add meat and cook until browned, turning to brown evenly.

2. Place onions in slow cooker, then top with meat. Add garlic, chipotle pepper, cilantro, seasoned salt, and juice. Cover and cook on low for 8 to 9 hours.

3. Remove meat and onions with a slotted spoon. Shred meat using two forks; stir together shredded meat and onion.

4. Spoon meat and fresh tomato salsa into the center of a warmed tortilla, then fold tortilla over filling. Serve immediately.

Fresh Tomato Salsa

8 Roma tomatoes, seeded and chopped

4 green onions, sliced

$\frac{1}{4}$ cup minced fresh cilantro

2 jalapeno peppers, seeded and minced

Juice of 2 limes

✦✦✦

1. Stir together all ingredients.
2. Serve immediately, or cover and refrigerate until serving time.

Easy Beef Burgundy

2 pounds beef stew meat

1 medium onion, chopped

1 can (4 ounces) sliced mushrooms, drained

1 package (1⅜ ounces) beef stew mix

Salt and pepper, to taste

1 cup water

¼ cup red wine

Hot cooked noodles

✦ ✦ ✦

1. Place beef stew meat, onion, and mushrooms in slow cooker.

2. Sprinkle with beef stew mix and season lightly to taste with salt and pepper. Add water and red wine, and stir to blend.

3. Cover and cook on low setting for 8 to 10 hours. Serve meat over hot cooked noodles.

Makes 6 to 8 servings.
Preparation time: 5 minutes
Cooking time: 8 to 10 hours
Slow cooker size: Medium;
   round

Favorite Taco Meat

TIPS

For extra quick preparation, cook ground beef when you have the time, then freeze in 1-pound packages. For this dish, for example, set two packages in the refrigerator the night before, then place the meat and remaining ingredients into the slow cooker the next morning.

For spicier taco meat, stir in another teaspoon of chili powder and add $\frac{1}{2}$ teaspoon of hot pepper sauce to meat.

No matter how you serve this meat, be sure to add all of the toppings. Sour cream, diced tomatoes, chopped green onions, sliced olives, shredded cheese, guacamole, sliced jalapeno peppers, and additional salsa are just a few of the available options.

Makes 8 servings.
Preparation time: 10 to
    15 minutes
Cooking time: 5 to 7 hours
Slow cooker size: Medium;
    round

Dinner will be on the table in almost no time. Use this seasoned meat as a filling for tacos, fold into warm flour tortillas, spoon over a taco salad, top nachos, fill enchiladas, or add to quesadillas.

2 pounds ground beef, browned and drained

$\frac{1}{2}$ cup chopped onion

3 cloves garlic, minced

2 teaspoons chili power

1 teaspoon cumin

$\frac{1}{2}$ cup salsa

✦✦✦

1. Combine all ingredients in slow cooker. Cover, and cook on low setting for 5 to 7 hours.

French Dip Roast Beef Sandwiches

1 medium onion, chopped

1 boneless beef rump roast, about 2½ to 3 pounds

1 tablespoon dry minced garlic

½ teaspoon dried thyme leaves

½ teaspoon dried basil leaves

½ teaspoon coarsely ground black pepper

1 can (10 ounces) beef consommé

Salt and pepper, to taste

8 to 10 crusty French rolls, split

✳ ✳ ✳

1. Place onion in slow cooker. Place beef on top of onion. Combine garlic, thyme, basil, and pepper; sprinkle over meat.

2. Pour consommé around beef. Cover, and cook on low setting for 9 to 10 hours.

3. Remove beef from drippings, and allow to stand for 10 to 15 minutes. Strain broth through sieve. Allow collected broth to stand for 10 to 15 minutes. Skim fat from broth. Taste broth and season, if desired, with salt and pepper.

4. Pour broth into individual soup cups or ramekins. Thinly slice beef. Pile beef onto buns, then serve with a cup of broth.

Makes 8 to 10 servings.
Preparation time: 5 minutes
Cooking time: 9 to 10 hours
Slow cooker size: Medium or large; round or oval

Hungarian Goulash

An easy-to-make dish that offers comfort and flavor! Who could ask for more?

1½ pounds round steak, cut into thick strips

3 tablespoons all-purpose flour

Salt and pepper, to taste

1 onion, chopped

1½ cups sliced mushrooms

1 can (14.5 ounces) beef broth

1 teaspoon dry minced garlic

1 teaspoon Hungarian paprika

1 teaspoon Worcestershire sauce

1 cup low-fat sour cream

3 cups egg noodles, cooked according to package directions and drained

✦✦✦

1. Place steak, flour, salt, and pepper in a zip-top bag. Shake to coat, then place in slow cooker.

2. Add onion, mushrooms, beef broth, garlic, paprika, and Worcestershire sauce. Cover, and cook on low for 7 to 9 hours.

3. Stir in sour cream until blended. Serve on top of noodles.

Makes 6 servings.
Preparation time: 10 to 15 minutes
Cooking time: 7 to 9 hours
Slow cooker size: Medium or large; round or oval

Italian Beef

TIPS

This meat is great for sandwiches. Top toasted bread with slices of Italian beef, thinly sliced tomatoes, sliced red onions, lettuce, and mayonnaise, or vinegar and oil.

> *Many think only of pot roast and potatoes made in the slow cooker, but this roast is flavorful and delightfully different. Serve it warm, with garlic mashed potatoes, or thinly sliced and served on sandwiches. In fact, the spicy blend on the roast will remind you of an Italian deli!*

2 teaspoons dry minced garlic

1½ teaspoons salt

1 teaspoon dried oregano leaves

1 teaspoon dried fennel seed

1 teaspoon anise seed

1 teaspoon dried basil leaves

½ teaspoon paprika

½ teaspoon pepper

1 boneless beef rump roast, about 3 pounds

1 can (8 ounces) tomato sauce

✦✦✦

1. Combine seasonings. Place meat in slow cooker and sprinkle on all sides with seasonings mix.
2. Drizzle tomato sauce over meat. Cover and cook on low setting for 8 to 9 hours.
3. Remove meat from drippings and allow to stand for 5 to 10 minutes. Thinly slice meat.

Makes 10 to 12 servings.
Preparation time: 5 to
   10 minutes
Cooking time: 8 to 9 hours
Slow cooker size: Medium or
   large; round or oval

MFK'S Prune Roast

TIPS

This roast is awesome when accompanied by mashed potatoes. If any of the roast is left over, you might serve it on toasted hamburger buns.

> M. F. K. Fisher was an inspirational food writer, and perhaps no one has ever written so passionately about eating and food. One of her recipes inspired this unique dish, and it is a privilege for us to acknowledge her influence on this recipe.

1 tablespoon vegetable oil

1 boneless beef rump roast, about 4 to 5 pounds

Salt and pepper, to taste

2 cups dried plums

$\frac{1}{2}$ cup water

$\frac{1}{3}$ cup cider vinegar

$\frac{2}{3}$ cup brown sugar

$\frac{1}{4}$ teaspoon ground cloves

1 teaspoon ground cinnamon

✦✦✦

1. Heat oil in heavy skillet and brown the roast on all sides. Salt and pepper meat generously, then place in slow cooker.

2. Add dried plums and water. Cover and cook on low for 8 to 10 hours.

3. Remove roast from slow cooker and shred using two forks; place on platter and cover with foil to keep warm.

4. Combine vinegar, brown sugar, cloves, and cinnamon in a saucepan. Add juices from roast. Simmer for 20 to 30 minutes, or until sauce is thickened. Pour over roast.

Makes 10 to 15 servings.
Preparation time: 15 to
    20 minutes
Cooking time: 8 to 10 hours
Slow cooker size: Medium or
    large; round or oval

Old-Fashioned Roast and Vegetables

> We have shared this recipe at cooking events and fairs all over the country, and it is a winner. Everyone loves how easy it is to prepare, as well as the rich flavor.

TIPS

This recipe is especially quick to prepare since the new potatoes are not peeled, and baby carrots are used. Feel free to substitute peeled potatoes, cut into 1-inch chunks, and sliced carrots if you wish.

To make cleanup faster, place potatoes and carrots in a zip-top bag, add oil, and then add seasoning; seal and shake to coat.

1 package (1 ounce) brown gravy mix
1 tablespoon dry minced garlic
1 teaspoon pepper
½ teaspoon onion salt
1 boneless beef rump roast, about 3 pounds
2 pounds new potatoes
1 cup baby carrots
2 to 3 tablespoons olive oil
½ cup water

✦ ✦ ✦

1. Combine brown gravy mix, dry minced garlic, pepper, and onion salt. Measure out 2 tablespoons of mixture and rub over roast. Place roast in slow cooker.

2. Toss potatoes and carrots in olive oil, then toss with remaining seasoning mix.

3. Place vegetables around roast and add water. Cover and cook on low for 8 to 10 hours.

Makes 6 to 8 servings.
Preparation time: 5 to
   10 minutes
Cooking time: 8 to 10 hours
Slow cooker size: Medium or
   large; round or oval

Pesto-Stuffed Round Steak

TIPS

For even more flavor, use your own, homemade Marinara Sauce (page 55) for this dish. If using frozen Marinara Sauce, thaw before pouring over the steak rolls.

These steak rolls should be cooked until tender, but do not overcook them or they can taste dry.

> *Pesto, with its delightful flavor of basil, garlic, and Parmesan cheese, permeates the steak. When cut into pinwheel-shaped slices and nestled on spaghetti, it looks elegant too!*

2 pounds boneless, beef round steak, cut about ½ inch thick

¼ cup basil pesto (see recipe, page 10)

½ cup chopped mushrooms

1 tablespoon olive oil

1 teaspoon dried basil leaves

¼ teaspoon salt

¼ teaspoon pepper

1 cup Marinara Sauce (see recipe, page 55)

Hot cooked spaghetti

Marinara Sauce, warmed

✳ ✳ ✳

1. Cut round steak into four serving pieces. Spread pesto on one side of each steak, then top with chopped mushrooms.

2. Roll steak in jelly-roll fashion. Tie with string, then place seam-side down in slow cooker.

3. Mix together olive oil, basil, salt, and pepper; brush evenly over the top of each steak roll. Drizzle 1 cup of Marinara Sauce over the top.

4. Cover and cook on low for 5 to 6 hours, or until meat is tender. Remove beef rolls and allow to stand for 5 to 10 minutes.

5. Slice meat 1 inch thick and arrange on top of spaghetti; top with additional warmed Marinara Sauce.

Makes 6 servings.
Preparation time: 10 to 15 minutes
Cooking time: 5 to 6 hours
Slow cooker size: Medium; round

Pizza in a Bowl

Planning a party for kids? This recipe fits the bill when you are in a time crunch and the youth group or Scout troop is coming to your home after a day filled with activities. Kids adore this one!

1 pound ground chuck

1 can (16 ounces) kidney beans, rinsed and drained

1 can (15 ounces) pizza sauce

1 can (14.5 ounces) diced tomatoes with basil, oregano, and garlic

1 can (6 ounces) tomato paste

1 cup water

1 package (3½ ounces) pepperoni, cut into fourths

2 teaspoons Italian seasoning

Hot cooked pasta, optional

1 cup shredded mozzarella cheese

1. Cook ground chuck in skillet over medium-high heat; drain and add to slow cooker.

2. Stir in remaining ingredients, except pasta and mozzarella cheese.

3. Cover and cook on low setting for 5 to 7 hours. Ladle over pasta, if desired. Sprinkle each serving with cheese.

Makes 4 to 6 servings.
Preparation time: 10 to
   15 minutes
Cooking time: 5 to 7 hours
Slow cooker size: Medium;
   round

Short Ribs with Merlot-Sage Sauce

TIPS

Be sure to bake the ribs in a hot oven until browned and crisp before placing them in the slow cooker.

Short ribs are not a tender cut of beef and need the slow simmering to become tender.

Short ribs are not as readily available year-round as some cuts at grocery-store meat counters. Go ahead and pick them up when you see them, package about 3 pounds together, and freeze until ready to use.

Makes 3 to 4 servings.
Preparation time: 20 to 30 minutes
Cooking time: 8 to 10 hours
Slow cooker size: Medium; round or oval

This recipe sounds elegant, but it is quite easy to prepare.

3 to 4 pounds beef short ribs
10 boiling onions, peeled
$\frac{1}{2}$ teaspoon salt
$\frac{1}{4}$ teaspoon pepper
1 teaspoon rubbed sage
$\frac{1}{2}$ cup beef broth
$\frac{1}{2}$ cup Merlot
1 tablespoon all-purpose flour
2 tablespoons cold water
1 tablespoon fresh minced sage leaves

✦ ✦ ✦

1. Place ribs on broiler pan; bake at 500° for 20 to 30 minutes, or until well browned. Drain and place in slow cooker.

2. Place onions around beef; season with salt, pepper, and sage. Pour broth and Merlot over meat. Cover and cook on low for 8 to 10 hours.

3. Remove ribs and onions and place on serving platter; keep warm. Ladle juices into small saucepan; skim excess fat from juices, if desired.

4. Blend together flour and water, stirring to form a smooth paste. Blend flour mixture into collected juices. Stir in sage.

5. Cook over medium heat, stirring constantly, until thickened and bubbly. Adjust seasonings with salt and pepper, if desired. Spoon sauce over meat.

Spicy Swiss Steak

TIPS

Bags of prepared, grated carrots are now available in the produce sections of many grocery stores. This is a quick way to get the thickly grated carrots for this recipe.

> Mention Swiss steak, and people may think of Mom's home-style cooking or an old-fashioned diner. Neither image applies this time! The flavor of this dish is fresh and fantastic.

¼ cup all-purpose flour

Salt and pepper, to taste

1½ pounds top sirloin, about 1-inch thick, cut into 6 pieces

1 tablespoon vegetable oil

1 onion, chopped

½ stalk celery, thinly sliced

¾ cup thickly grated carrot

1 jalapeno, seeded and finely chopped

½ cup picante sauce

¼ cup ketchup

1 tablespoon cider vinegar

Hot cooked noodles

✳ ✳ ✳

1. Place flour, salt, and pepper in zip-top bag. Add beef in batches, and toss to coat generously with flour.

2. Heat oil in large skillet over medium-high heat; add beef and sear on both sides. Place meat in slow cooker.

3. Add onion to skillet and cook, stirring frequently, for 5 to 6 minutes, or until tender and golden.

4. Combine the onion with the remaining ingredients, except noodles, and ladle over each piece of beef.

5. Cover and cook on low setting for 7 to 8 hours. Serve over hot cooked noodles.

Makes 6 servings.
Preparation time: 10 to 15 minutes
Cooking time: 7 to 8 hours
Slow cooker size: Medium or large; round or oval

Stuffed Flank Steak

TIPS

Packaged stuffing cubes makes this dish quick to prepare.

1 ½ cups cubed bread stuffing

¾ cup sliced mushrooms

3 tablespoons water

2 tablespoons melted butter or margarine

2 tablespoons grated Parmesan cheese

1 ½ to 2 pounds beef flank steak

¼ cup dry red wine

½ cup beef broth

2 green onions, thinly sliced

⅓ cup currant jelly

✦✦✦

1. Combine stuffing, mushrooms, water, butter, and Parmesan cheese.

2. Cut flank steak in half; place half in slow cooker. Carefully spoon stuffing over the top of the steak, mounding in the center. Place remaining half of flank steak over stuffing, covering it.

3. Pour wine and broth over steak. Sprinkle with green onions. Cover and cook on low setting for 6 to 8 hours.

4. Just before serving, heat jelly in microwave oven on high power (100%) for 30 seconds to 1 minute, or until melted; brush over steak.

5. Cook on high setting for 5 minutes. To serve, lift up both pieces of meat with a spatula, with the stuffing in between, and place on serving platter. Slice into serving portions.

Makes 6 servings.
Preparation time: 5 to
   10 minutes
Cooking time: 6 to 8 hours
Slow cooker size: Medium or
   large; round or oval

Stuffed Italian Steak

1 ½ pounds boneless beef round steak, cut about ½ inch thick

½ teaspoon garlic salt

¼ teaspoon coarsely ground pepper

1 medium carrot, grated coarsely

1 medium zucchini squash, thinly sliced

⅔ cup shredded mozzarella cheese

2 cups Marinara Sauce (see recipe, page 55)

✦✦✦

1. Pound round steak until very thin. Sprinkle with garlic salt and pepper.

2. Sprinkle carrots over steak, then layer zucchini on carrots. Sprinkle cheese down the center.

3. Roll steak from the long side, in jelly-roll fashion, and secure with cooking twine.

4. Place in slow cooker and cover with Marinara Sauce. Cover and cook on low setting for 6 to 8 hours. Remove string, and slice into thick slices to serve.

Makes 6 servings.
Preparation time: 10 to 15 minutes
Cooking time: 6 to 8 hours
Slow cooker size: Medium or large; round or oval

PORK

BBQ Baby Back Ribs

1 tablespoon vegetable oil

1 cup chopped onion

3 cloves garlic, minced

1 $\frac{1}{2}$ cups ketchup

$\frac{1}{4}$ cup molasses

2 tablespoons cider vinegar

Dash of hot pepper sauce

1 tablespoon brown sugar

$\frac{1}{2}$ teaspoon salt

1 teaspoon pepper

2 racks baby back ribs

✦✦✦

1. Heat oil in skillet; add onion and garlic. Cook for 5 minutes, until tender and golden. Add ketchup, molasses, vinegar, hot sauce, brown sugar, salt, and pepper.

2. Place ribs on broiler pan; broil for approximately 8 minutes on each side, or until browned.

3. Place in slow cooker. Pour sauce over ribs. Cover and cook on low setting for 6 to 8 hours.

Makes 4 servings.
Preparation time: 15 to
   20 minutes
Cooking time: 6 to 8 hours
Slow cooker size: Medium or
   large; oval

Cider-Glazed Pork Chops

¼ cup all-purpose flour

½ teaspoon salt

½ teaspoon ground nutmeg

¼ teaspoon pepper

4 to 6 boneless pork chops, about 1½ pounds, cut ¾ to 1 inch thick

2 tablespoons vegetable oil

1 medium onion, cut into 8 to 10 thin wedges

2 medium sweet potatoes, peeled and cut into quarters

⅓ cup dried cranberries

1 large firm, tart apple, such as a Rome, cored and sliced into 3 rings

2 tablespoons lemon juice

½ cup apple cider

✳✳✳

1. Combine flour, salt, nutmeg, and pepper. Place chops in zip-top bag and sprinkle with flour mixture. Shake to coat pork chops evenly.

2. Heat oil in large skillet over medium-high heat. Cook chops until well browned on each side; drain.

3. Place onions and sweet potatoes in slow cooker. Arrange chops over vegetables. Sprinkle with cranberries.

4. Coat apple slices in lemon juice, and arrange over top of chops. Add apple cider.

5. Cover and cook on low setting for 5 to 7 hours, or until pork chops are tender. To serve, arrange chops, apples, and vegetables on large platter. Spoon some of the collected drippings over chops.

Makes 4 to 6 servings.
Preparation time: 15 minutes
Cooking time: 5 to 7 hours
Slow cooker size: Medium or large; round or oval

Curried Pork with Couscous

Though some families may not think about serving a curried dish, flavored with cayenne, mint, and currants, this recipe is a winner! Give it a try.

TIPS

Couscous is granular wheat or semolina. It rounds out a meal very quickly and cooks almost instantly in hot liquids. Add it to the juices in this recipe just before serving.

4 boneless pork chops or about 1¾ pounds of boneless pork loin

2 tablespoons vegetable oil

½ cup chopped onion

2 cloves garlic, minced

⅓ cup pine nuts

¼ teaspoon salt

¼ teaspoon cayenne pepper

1 teaspoon curry powder

¼ cup currants

1 tablespoon lemon juice

1 can (14.5 ounces) chicken broth

2 tablespoons minced parsley

1 tablespoon minced fresh mint

1 cup couscous

✦✦✦

1. Cut pork into ¾-inch cubes. Heat oil in large skillet over medium-high heat. Add pork cubes and cook, stirring frequently, until browned.

2. Remove pork with slotted spoon, leaving drippings, and place pork in slow cooker.

3. Add onion and garlic to drippings and cook, stirring frequently, until onion is transparent. Add pine nuts and cook, stirring frequently, for 1 to 2 minutes, or until pine nuts are golden.

Makes 4 to 6 servings.
Preparation time: 15 minutes
Cooking time: 4 to 6 hours
Slow cooker size: Medium or large; round or oval

4. Stir onion-pine nut mixture into pork. Add seasonings, currants, lemon juice, and broth. Cover and cook on low setting for 4 to 6 hours.

5. Stir parsley, mint, and couscous into pork. Cover, turn slow cooker off, and allow to stand for 5 minutes. Fluff couscous with fork.

Kielbasa, Cabbage, and Kraut

½ cup chopped onion

1 pound kielbasa, cut into 1-inch pieces

6 to 8 new potatoes

1 cup beef broth

1 tablespoon brown sugar

½ teaspoon caraway seed

½ teaspoon salt

½ teaspoon pepper

4 cups thinly sliced cabbage

2 cups refrigerated sauerkraut, drained

✦✦✦

1. Place onion, kielbasa, potatoes, broth, brown sugar, caraway seed, salt, and pepper in slow cooker. Cover and cook on low setting for 6 to 7 hours.

2. Turn to high setting and add cabbage and kraut; cover and cook for 1 hour.

Makes 4 servings.
Preparation time: 10 to
   15 minutes
Cooking time: 7 to 8 hours
Slow cooker size: Medium;
   round or oval

Pork Roast Florentine

TIPS

Two readily available convenience foods make this dish even easier to prepare. Washed, ready-to-serve spinach leaves are great. This is one time when the ready-to-serve, or packaged, fully-cooked bacon is even better than bacon strips you might cook at home. This bacon, while cooked, is not crisp and conforms to the shape of the roast.

1 boneless pork loin roast, about 2 to 2½ pounds
2 tablespoons vegetable oil
3 tablespoons cream cheese, softened
2 cloves garlic, minced
1 teaspoon Italian seasoning
¼ teaspoon salt
¼ teaspoon pepper
¾ cup fresh, trimmed spinach leaves
1 tablespoon balsamic vinegar
4 slices ready-to-serve bacon

1. Cut roast horizontally, through center, not quite cutting through one side. Fold open to lay flat.

2. Heat oil in large skillet over medium-high heat. Place roast, cut-side up, in skillet; cook until outside is well browned.

3. Remove from skillet and place on cutting board. (Do not brown the cut side of the meat.) Spread cream cheese over one side of cut surface of meat. Sprinkle cream cheese with garlic, Italian seasoning, salt, and pepper.

4. Arrange spinach leaves over meat. Fold meat over; tie shut with string.

5. Place in slow cooker, and drizzle with balsamic vinegar. Arrange bacon over roast. Cover and cook on low setting for 7 to 8 hours.

Makes 6 to 8 servings.
Preparation time: 10 to
   15 minutes
Cooking time: 7 to 8 hours
Slow cooker size: Medium or
   large; round or oval

Provencal Pork Chops

> *These thick-cut chops are thicker than often selected—and that is one thing that makes this dish so very good. Good enough, in fact, for company.*

2 tablespoons all-purpose flour

1 teaspoon dried thyme leaves

½ teaspoon salt

¼ teaspoon pepper

4 thick-cut, bone-in pork chops, about 3 pounds, each cut 1½ inches thick

4 slices bacon

6 cloves garlic, minced

1 onion, cut into 8 wedges

8 ounces mushrooms, sliced

½ teaspoon dried thyme leaves

¼ teaspoon salt

¼ cup white wine

2 tablespoons water

✴ ✴ ✴

1. Combine flour, 1 teaspoon of thyme, ½ teaspoon of salt, and ¼ teaspoon of pepper. Sprinkle over each side of chops, coating evenly.

2. Fry bacon in large skillet over medium heat, until crisp. Remove bacon, leaving drippings.

3. Increase heat to medium-high. Add chops, and cook just until browned on each side. Remove chops, leaving drippings, and place in slow cooker.

Makes 4 to 6 servings.
Preparation time: 15 to 20 minutes
Cooking time: 5 to 6 hours
Slow cooker size: Large; oval

4. Add garlic and onion to skillet and cook, stirring frequently, until golden. Add mushrooms, and sauté for 2 to 3 minutes. (Add a small amount of vegetable oil if there are not enough bacon drippings.) Season vegetables with ½ teaspoon of thyme and ½ teaspoon of salt.

5. Drain and spoon into slow cooker. Add wine and water. Crumble bacon and sprinkle over all. Cover and cook on low setting for 5 to 6 hours.

6. Just before serving, ladle out 1 cup of drippings. Pour through strainer into small saucepan. Return any collected bacon and onions to slow cooker. Heat broth over high heat, uncovered, until reduced by about half.

7. Arrange chops with vegetables in deep platter. Drizzle with reduced broth.

Shredded Pork Tortillas

1 package (1 ¼ ounces) taco seasoning mix

1 boneless pork shoulder roast, about 3 to 3 ½ pounds

2 cloves garlic, crushed

1 cup chicken broth

1 jar (8 ounces) picante sauce

12 flour tortillas

Sour cream, chopped green onions, shredded cheddar cheese, salsa, or chopped avocados, for topping

✳ ✳ ✳

1. Rub taco seasoning mix into pork and place in slow cooker. Combine garlic, broth, and picante sauce; pour around roast. Cover and cook on low setting for 8 to 10 hours.

2. Remove meat from slow cooker, reserving drippings. Shred meat with forks and return to cooker; stir well to combine with drippings.

3. To serve, spoon shredded meat into warmed tortillas and roll up. Top as desired with sour cream, chopped green onions, cheese, salsa, and avocado.

Makes 8 to 12 servings.
Preparation time: 5 minutes
Cooking time: 8 to 10 hours
Slow cooker size: Medium or large; round or oval

Slow-Roasted Cola Pork

TIPS

Freeze leftovers for a busy nighttime dinner solution.

Pork butt is less expensive and perfect for slow cooking because the connective tissue will dissolve. It may not be on display at the meat counter, but ask the butcher, as he or she will usually have this cut on hand.

1 boneless pork butt roast, about 4 to 7 pou

$\frac{1}{4}$ cup Worcestershire sauce

$\frac{3}{4}$ cup brown sugar

1 cup cola-flavored carbonated beverage

$\frac{1}{2}$ teaspoon salt

✦✦✦

1. Place pork in slow cooker. Drizzle evenly with Worcestershire sauce. Press brown sugar on top and sides of pork.

2. Carefully pour cola around pork. Cover and cook on low setting for 9 to 11 hours.

3. Remove meat from slow cooker. Shred meat using two forks and return it to the slow cooker; add salt and heat on high setting for 30 minutes.

Makes 12 to 14 servings.
Preparation time: 5 minutes
Cooking time: $9\frac{1}{2}$ to
    $11\frac{1}{2}$ hours
Slow cooker size: Medium or
    large; round or oval

Southwestern Pork Loin with Spicy Cilantro Pesto

This roast is wonderful, as is—delightfully seasoned with lime, garlic, and cumin. Yet, for a truly amazing flavor, add a spoonful of cilantro pesto and the flavor will just soar.

1 boneless pork loin roast, about 3 to 3½ pounds

1 teaspoon grated lime zest

Juice of 1 lime

1 tablespoon dry minced garlic

1 teaspoon cumin

1 teaspoon pepper

½ teaspoon salt

1 tablespoon vegetable oil

Spicy Cilantro Pesto (recipe following)

✦ ✦ ✦

1. Place roast in slow cooker. Make a paste of lime zest, lime juice, garlic, cumin, pepper, salt and oil; rub into surface of roast. Cover and cook on low setting for 7 to 8 hours.

2. Remove from slow cooker and allow to stand for 10 to 15 minutes. Thinly slice and serve with cilantro pesto.

Makes 8 to 10 servings.
Preparation time: 5 minutes
Cooking time: 7 to 8 hours
Slow cooker size: Medium or large; round or oval

Spicy Cilantro Pesto

¼ cup grated Parmesan cheese

2 cloves garlic, halved

1 jalapeno pepper, seeded

⅓ cup toasted pine nuts

¼ cup cilantro leaves

¼ teaspoon salt

¼ cup olive oil

+ + +

1. Place Parmesan cheese, garlic, jalapeno, pine nuts, cilantro, and salt in work bowl of food processor. Pulse to finely chop.

2. With motor running, drizzle in olive oil and process until well blended.

Spicy Smoked Sausage Hoagies

> *This is the perfect dish to serve when a crowd is gathering to watch the big game on television. It is quick to fix and just spicy enough to be really great!*

TIPS

Toasting the buns helps to ensure that the sandwiches don't get too soggy.

If desired, sprinkle the hot sausages and vegetables with shredded Monterey Jack, Pepper Jack, or cheddar cheese.

1 large onion, thinly sliced

1 red pepper, thinly sliced

1 green pepper, thinly sliced

2 pounds smoked sausage, sliced into 1-inch pieces

2 teaspoons chili powder

1 can (14.5 ounces) diced tomatoes

1 can (10 ounces) diced tomatoes and green chiles

8 hoagie or sandwich buns, split and toasted

1. Place onion, red pepper, and green pepper in slow cooker. Top with sliced sausage.

2. Sprinkle with chili powder, then top with tomatoes, and tomatoes and green chilies. Cover and cook on low setting for 6 to 7 hours, or on high setting for 2 to 3 hours.

3. Using a slotted spoon, spoon sausages and peppers out of liquid and place on toasted buns.

Makes 8 servings.
Preparation time: 5 minutes
Cooking time: 6 to 7 hours on low, or 2 to 3 hours on high
Slow cooker size: Medium or large; round or oval

CHICKEN AND POULTRY

Biscuit-Topped Chicken Potpie

Comforting—there is just no other way to describe a creamy, rich potpie.

TIPS

Be sure to use a slow cooker with a removable stoneware or crockery bowl. (Never place the electrical heating base of the slow cooker in the oven.) If your slow cooker has a stationary bowl, spoon the cooked chicken mixture into a 3-quart casserole dish. Top with teaspoonfuls of biscuit dough and bake as directed.

¼ cup all-purpose flour

½ teaspoon salt

¼ teaspoon pepper

1½ pounds boneless, skinless chicken breasts, cut into ¾-inch cubes

2 tablespoons vegetable oil

1 medium onion, chopped

2 cups sliced mushrooms

2 medium carrots, sliced

2 stalks celery, sliced

¼ cup dry sherry

½ cup chicken broth

½ teaspoon dry mustard

¼ teaspoon dried tarragon leaves

1 cup frozen peas

½ cup shredded Parmesan cheese

¼ cup heavy whipping cream

1⅔ cups baking mix

2 tablespoons shredded Parmesan cheese

½ cup milk

✦✦✦

Makes 4 to 6 servings.
Preparation time: 15 minutes
Cooking time: 5 to 6 hours
Baking time: 15 to
    20 minutes
Slow cooker size: Medium;
    round or oval

1. Combine flour, salt, and pepper in zip-top bag. Add chicken breast cubes; seal and shake to coat.

2. Heat oil in large skillet over medium-high heat. Add chicken and cook, stirring frequently, until chicken is golden. (Add an additional 1 to 2 tablespoons of oil to the skillet, if necessary.)

3. Remove chicken and place in slow cooker. Add onion to the skillet and cook, stirring frequently, for 2 to 3 minutes. Add mushrooms and continue cooking until onions are tender. Spoon into slow cooker.

4. Add carrots, celery, sherry, broth, mustard, and tarragon to slow cooker; stir well. Cover and cook on low setting for 5 to 6 hours.

5. Turn slow cooker to high setting. Stir in peas, ½ cup of shredded Parmesan, and cream. Cover and allow to cook just while preparing biscuit topping.

6. Preheat oven to 425°. Stir together baking mix, 2 tablespoons of Parmesan cheese, and milk. Drop by teaspoonfuls over chicken mixture.

7. Remove crockery from slow cooker base and place it in a hot oven. Bake for 15 to 20 minutes, or until biscuits are golden.

Chicken and Artichokes

Scrumptious. There is just no other way to describe the flavor combination of chicken and artichokes. Enjoy!

¼ cup all-purpose flour

½ teaspoon salt

¼ teaspoon pepper

6 boneless, skinless chicken breast halves

¼ cup olive oil

1 cup chopped onion

2 cloves garlic, minced

1 jar (6 ounces) marinated artichoke hearts, drained and quartered

½ cup chicken broth

¼ cup sour cream

½ cup shredded Parmesan cheese

Hot cooked pasta, optional

1. Combine flour, salt, and pepper in zip-top bag; add chicken, seal and shake to coat well.

2. Heat about 2 tablespoons of oil in large skillet over medium-high heat. Brown three pieces of chicken; remove chicken from skillet, leaving drippings.

3. Place chicken in slow cooker. Repeat with remaining oil and three pieces of chicken.

4. Add onion to skillet and cook until golden; spoon into slow cooker. Add garlic, artichokes, and broth to slow cooker. Cover and cook on low setting for 3 to 4 hours.

Makes 6 servings.
Preparation time: 30 minutes
Cooking time: 3 to 4 hours
Slow cooker size: Medium or large; round or oval

5. Remove chicken and artichokes, and place on a deep platter. Ladle collected broth through a strainer into a saucepan (broth should measure about 1 cup). Heat to a boil; boil uncovered for about 3 minutes.

6. Whisk in sour cream, stirring until smooth. Pour sauce over chicken. Sprinkle with cheese. Serve with pasta, if desired.

Chicken and Rice

Makes 6 servings.
Preparation time: 10 to 15 minutes
Cooking time: 5 to 6 hours
Slow cooker size: Medium or large; round or oval

½ cup uncooked brown rice

⅔ cup uncooked parboiled white rice

3 tablespoons butter or margarine

½ cup chopped onion

1 can (4 ounces) sliced mushrooms, drained

½ teaspoon dried thyme leaves

½ teaspoon rubbed sage

½ teaspoon dry minced garlic

½ teaspoon salt

¼ teaspoon pepper

6 chicken thighs

1 can (10.5 ounces) beef consommé

2 tablespoons Worcestershire sauce

½ teaspoon paprika

½ teaspoon dried thyme leaves

✳✳✳

1. Combine brown rice, rice and butter in skillet. Cook over medium-high heat, stirring occasionally, until rice is golden brown.

2. Remove from heat and stir in onions, mushrooms, thyme, sage, garlic, salt, and pepper.

3. Pour rice mixture into slow cooker. Arrange chicken over rice mixture. Pour broth over chicken, then drizzle Worcestershire sauce over chicken.

4. Combine paprika and remaining ½ teaspoon of thyme; sprinkle over chicken. Cover and cook on low setting for 5 to 6 hours.

Chicken Con Queso

A *fantastic chicken dish, ready-made for chicken nachos, tostados, or enchiladas.*

TIPS

Quesadilla: Lightly brush a flour tortilla with melted butter. Place tortilla, buttered-side down, on a preheated contact grill. Spoon about ¼ cup of chicken mixture over the tortilla. Top with a second tortilla, buttered-side up. Grill for 1 to 2 minutes, or until golden. Repeat with five other quesadillas.

Tostados: Fry six corn tortillas in hot vegetable oil in a skillet over medium-high heat until crisp. Drain and place on a baking sheet. Spoon chicken con queso over crisp tortillas. Bake at 350° for 5 minutes. Top, if desired, with pico de gallo, guacamole, sour cream, or salsa.

Enchiladas: Warm six flour tortillas, about 8 inches in diameter. Spoon about 2 tablespoons of chicken con

1 ½ to 2 pounds boneless, skinless chicken breast halves

1 jalapeno pepper, seeded and diced

1 clove garlic, minced

½ teaspoon ground cumin

¼ teaspoon salt

¼ teaspoon pepper

1 cup chicken broth

1 can (4 ounces) chopped green chiles, drained

1 cup shredded Monterey Jack cheese

½ cup chunky salsa

¼ cup sour cream

2 tablespoons minced cilantro

1. Place chicken breasts, jalapeno pepper, and garlic in slow cooker. Season with cumin, salt, and pepper. Pour chicken broth over chicken. Cover and cook on low for 3 to 4 hours.

2. Remove chicken, and drain. Shred chicken and return it to slow cooker.

3. Add green chiles, cheese, salsa, sour cream, and cilantro; stir well. Cover and cook on low for 30 minutes.

(continues on next page)

queso into the center of each tortilla. Fold tortilla over filling, then arrange in a 12-x-8-inch baking dish sprayed with nonstick vegetable coating. Spoon any remaining chicken con queso over the center of the tortillas. Sprinkle with additional cheese, if desired. Cover and bake at 350° for 20 to 30 minutes, or until hot and bubbly.

Nachos: Arrange tortilla chips on a heatproof baking dish. Spoon chicken con queso over chips. Sprinkle with additional cheese, if desired. Bake at 400° for about 5 minutes, or until cheese is melted. Top, as desired, with guacamole, sour cream, sliced, pickled jalapeno peppers, or other favorite toppings.

Makes 6 servings.
Preparation time: 5 minutes
Cooking time: $3\frac{1}{2}$ to
   $4\frac{1}{2}$ hours
Slow cooker size: Medium;
   round

Family-Style Turkey and Dressing with Savory Sage Sauce

No need to serve turkey and dressing only at Thanksgiving. The slow cooker makes it easy, and, best of all, you won't be eating leftovers for a week.

2 tablespoons vegetable oil

1 split, bone-in turkey breast, about 2½ to 3 pounds

1 medium onion, cut into 8 wedges

1 stalk celery, cut into 2-inch pieces

1 teaspoon dry minced garlic

1 teaspoon dried thyme leaves

1 tablespoon rubbed sage

½ teaspoon salt

½ teaspoon pepper

½ cup water

1 tart apple, peeled, cored, and finely chopped

1 package (6 ounces) turkey-flavored stuffing mix

2 tablespoons butter

2 tablespoons all-purpose flour

½ teaspoon rubbed sage

Salt and pepper, to taste

✳ ✳ ✳

1. Heat oil in large, heavy skillet over medium-high heat. Cook turkey, skin-side down, in skillet until well browned.

2. Place turkey, skin-side up, in slow cooker. Place onion wedges and celery around turkey.

3. Combine garlic, thyme, sage, salt, and pepper, and sprinkle over turkey breast. Drizzle water around turkey. Cover and cook on low setting for 8 to 10 hours.

(continues on next page)

Makes 6 servings.
Preparation time: 10 to 15 minutes
Cooking time: 8¼ to 10 hours 20 minutes
Slow cooker size: Large; oval

4. Remove turkey to a serving platter; cover and keep warm. Remove vegetables with a slotted spoon, and set aside to partially cool. Strain and reserve broth.

5. Place chopped apple in the slow cooker. Add 1½ cups of reserved broth. Cover and cook on high setting for about 10 minutes.

6. When cooked vegetables are cool enough to handle, chop ½ cup each of onion and celery; stir into apple. Stir in stuffing mix. Cover and allow to cook for 5 to 10 minutes.

7. Meanwhile, melt butter in a small saucepan. Stir in flour, blending to form a smooth paste. Cook for 1 to 2 minutes. Measure remaining broth; add water, if necessary, to equal 1 cup.

8. Gradually stir broth into saucepan, blending until smooth. Stir in sage. Cook, stirring constantly, until thickened and bubbly. Stir in salt and pepper, to taste.

9. Slice turkey and accompany with dressing. Serve with sage sauce.

Garlic-Roasted Chicken

While this is great when eaten hot and tasty right out of the slow cooker, you might want to use the leftover chicken in your favorite chicken salad. The garlic flavor will add depth and richness to even the most basic chicken salad sandwich.

1 whole broiler-fryer chicken, about 3 ½ pounds

1 tablespoon olive oil

2 sprigs rosemary

20 cloves garlic, peeled

1 teaspoon salt

½ teaspoon pepper

¼ cup dry white wine

2 tablespoons lemon juice

1. Rub chicken with olive oil. Place rosemary and 10 cloves of garlic in chicken cavity.

2. Place chicken in slow cooker. Sprinkle remaining garlic around chicken. Sprinkle salt and pepper over chicken.

3. Drizzle with wine and lemon juice. Cover and cook on low setting for 8 to 9 hours.

Makes 4 to 6 servings.
Preparation time: 10 minutes
Cooking time: 8 to 9 hours
Slow cooker size: Medium or large; round or oval

Lemon Chicken with Bacon and Rosemary

1 tablespoon olive oil

¼ cup all-purpose flour

Salt and pepper, to taste

4 boneless, skinless chicken breast halves

2 teaspoons dry minced garlic

2 teaspoons dried rosemary leaves, crushed

¼ teaspoon crushed red pepper flakes

1 cup chicken broth

2 tablespoons freshly squeezed lemon juice

4 slices bacon

✴✴✴

1. Heat oil in large skillet over medium-high heat. Combine flour, salt, and pepper in zip-top bag; add chicken, and shake to coat generously.

2. Place chicken in hot skillet and brown quickly on both sides.

3. Place chicken in slow cooker. Sprinkle with garlic, rosemary, and red pepper flakes. Pour chicken broth around chicken. Cover and cook on low setting for 3 to 4 hours.

4. Transfer chicken to platter and cover with foil; set aside. Pour drippings into saucepan. Add lemon juice and simmer until slightly thickened, about 5 minutes.

5. Place bacon on microwave-safe plate and microwave on high power (100%) for 3 to 4 minutes, or until crisp. Pour sauce over chicken, and sprinkle with crumbled bacon.

Makes 4 servings.
Preparation time: 15 minutes
Cooking time: 3 to 4 hours
Slow cooker size: Medium or large; round or oval

Tarragon-Smothered Chicken

8 chicken thighs

2 tablespoons butter

10 boiling onions, halved

8 ounces mushrooms, sliced

1 teaspoon salt

¼ teaspoon pepper

½ teaspoon paprika

2 teaspoons dried tarragon leaves

3 tablespoons white wine

3 tablespoons sour cream

½ teaspoon dried tarragon leaves

2 tablespoons all-purpose flour

2 tablespoons cold water

Hot cooked rice

✦ ✦ ✦

1. Brown chicken thighs in butter in large skillet over medium-high heat; place in slow cooker.

2. Add onions and mushrooms to skillet and cook until golden brown, stirring frequently.

3. Transfer vegetables to slow cooker and sprinkle all with salt, pepper, paprika, and tarragon. Drizzle with wine. Cover and cook on low setting for 6 to 7 hours.

4. With a slotted spoon, remove chicken thighs and vegetables to a deep serving platter and keep warm.

5. Turn slow cooker to high setting. Stir sour cream and ½ teaspoon of tarragon into drippings, blending until smooth. Blend together flour and water, stirring until smooth. Stir flour mixture into drippings.

6. Cover and cook on high for 20 minutes. Spoon sauce over chicken and serve with rice.

Makes 8 servings.
Preparation time: 15 minutes
Cooking time: 6 to 7 hours
Slow cooker size: Medium or large; oval

Turkey Casserole with Cornbread Topping

TIPS

Be sure to use a slow cooker with a removable stoneware or crockery bowl. (Never place the electrical heating base of the slow cooker in the oven.) If your slow cooker has a stationary bowl, spoon the cooked turkey mixture into a 3-quart casserole dish. Top with cornbread and bake as directed.

Makes 6 to 8 servings.
Preparation time: 20 minutes
Cooking time: $5\frac{1}{4}$ to 6 hours 20 minutes
Baking time: 15 to 20 minutes
Slow cooker size: Medium or large; round or oval

2 tablespoons all-purpose flour

$\frac{1}{2}$ teaspoon salt

$\frac{1}{4}$ teaspoon pepper

$\frac{1}{4}$ teaspoon cayenne pepper

$1\frac{1}{4}$ pounds boneless, skinless turkey breast, cut into 1-inch cubes

2 tablespoons vegetable oil

8 ounces fresh mushrooms, sliced

1 medium onion, chopped

2 carrots, sliced

1 stalk celery, sliced

1 medium potato, peeled and chopped

2 cloves garlic, minced

2 teaspoons poultry seasoning

1 can (14.5 ounces) chicken broth

2 tablespoons dry sherry

1 cup frozen mixed vegetables

2 tablespoons all-purpose flour

3 tablespoons cold water

Corn bread:

$\frac{1}{2}$ cup all-purpose flour

$\frac{1}{2}$ cup yellow cornmeal

$1\frac{1}{2}$ teaspoons baking powder

$\frac{1}{4}$ teaspoon salt

1 egg, lightly beaten

$\frac{1}{3}$ cup milk

2 tablespoons shortening

✦✦✦

1. Combine flour, salt, pepper, and cayenne in zip-top bag; add turkey, then seal and shake to coat well.

2. Heat oil in large skillet over medium-high heat. Add turkey and cook until lightly browned, stirring frequently.

3. Using a slotted spoon, spoon turkey into slow cooker, leaving drippings. (Add a small amount of additional oil, if necessary.) Add mushrooms and onion, and cook, stirring frequently, until tender.

4. Place in slow cooker and add carrots, celery, potato, garlic, poultry seasoning, broth, and sherry. Cover and cook on low setting for 5 to 6 hours.

5. Turn slow cooker to high setting. Stir in frozen mixed vegetables.

6. Stir together 2 tablespoons of flour and cold water until it forms a smooth paste. Stir flour mixture into slow cooker. Cover and cook on high for 15 to 20 minutes.

7. Preheat oven to 400°. Stir together flour, cornmeal, baking powder, and salt in a mixing bowl.

8. In a small mixing bowl, combine egg, milk, and shortening. Stir milk mixture into dry ingredients, stirring just until combined. Pour cornmeal batter over turkey mixture.

9. Remove crockery from slow cooker base and place in hot oven. Bake for 15 to 20 minutes, or until cornbread is set and golden brown. Allow to stand for 5 to 10 minutes before serving.

VEGETABLES AND SIDES

Bavarian Sauerkraut

TIPS

For a main-dish dinner, feel free to add smoked kielbasa.

This sauerkraut is great on Rueben sandwiches, and you can toast the sandwiches on your contact grill.

2 pounds fresh sauerkraut, rinsed and drained

1 large yellow onion, chopped fine

2 tablespoons brown sugar

2 tablespoons butter or margarine

2 teaspoons caraway seeds

✦✦✦

1. Combine all ingredients in slow cooker. Cover and cook on low setting for 4 to 6 hours.

Makes 6 to 8 servings.
Preparation time: 5 minutes
Cooking time: 4 to 6 hours
Slow cooker size: Medium;
  round or oval

Creamy Corn

Our lives our woven with many food traditions. This recipe is always served at Roxanne's Thanksgiving dinner table. It frees the stovetop and oven for other dishes on a day that is filled with blessings.

2 packages (16 ounces each) frozen corn

1 package (8 ounces) cream cheese, cut into pieces

⅓ cup butter, cut into pieces

2 tablespoons sugar

2 tablespoons water

1. Place all ingredients in slow cooker. Toss lightly to combine.

2. Cover and cook on high setting for 30 minutes, then turn to low setting for 2 to 3 hours.

Makes 8 servings.
Preparation time: 5 minutes
Cooking time: 2½ to 3½ hours
Slow cooker size: Medium; round

Garlic Smashed Potatoes

> *Oh, so yummy and comforting, these potatoes are the perfect way to complete so many meals.*

TIPS

Coating the potatoes evenly in the salted water before cooking helps prevent darkening. However, this is one recipe that is best prepared just before you turn on the slow cooker; do not try to peel potatoes the night before.

Russet potatoes are a great choice to use for this recipe. However, you may enjoy trying finger potatoes, Yukon gold potatoes, or new potatoes. You can choose to leave the skin on these potatoes when you cook them; the cooked skin adds flavor and nutrition, but it may also darken and will add more texture.

Top these tasty potatoes with additional toppings, such as sliced green onion or crisp, crumbled bacon, if desired.

Makes 8 to 10 servings.
Preparation time: 10 minutes
Cooking time: 8 to 10 hours
Slow cooker size: Medium or
  large; round or oval

3 pounds potatoes, peeled and cut into $\frac{1}{2}$-inch cubes

1 teaspoon salt

$\frac{1}{2}$ teaspoon pepper

$\frac{1}{2}$ cup water

4 cloves garlic, minced

3 tablespoons chopped onion

2 tablespoons butter or margarine

2 to 3 tablespoons milk

$\frac{1}{3}$ cup sour cream

$\frac{3}{4}$ cup shredded cheddar cheese

✦✦✦

1. Place potatoes in slow cooker. Sprinkle with salt and pepper, and add water; toss to coat potatoes in salted water.

2. Add garlic, onion, and butter. Cover and cook on low setting for 8 to 10 hours, or until potatoes are very tender.

3. Add milk and sour cream; mash with potato masher while still in slow cooker. (Potatoes will still be lumpy.) Stir in cheese.

Home-Style Applesauce

Each fall, get your slow cooker simmering with nature's bounty. This will be a family tradition that is sure to please.

10 large cooking apples, such as Jonathan's

½ cup water

1½ teaspoons cinnamon

1 cup sugar

✦✦✦

1. Peel, core, and slice apples into bite-size pieces. Place in slow cooker along with remaining ingredients.
2. Cover and cook on low setting for 8 to 10 hours. Serve warm or chilled.

Makes 6 to 8 servings
Preparation time: 20 minutes
Cooking time: 8 to 10 hours
Slow cooker size: Medium or
  large; round or oval

Roasted Beet Salad

4 medium beets, scrubbed, trimmed, and peeled

1 tablespoon extra virgin olive oil

Kosher salt and freshly ground pepper to taste

1 large shallot, thinly sliced

1 tablespoon extra virgin olive oil

2 tablespoons prepared horseradish

1 tablespoon Dijon mustard

$\frac{1}{4}$ cup heavy whipping cream

1 teaspoon freshly squeezed lemon juice

1 tablespoon extra virgin olive oil

2 teaspoons freshly squeezed lemon juice

1 tablespoon extra virgin olive oil

10 cups lettuce, such as leaf, oak leaf, red leaf, and spring mix combination

$\frac{2}{3}$ cup walnut pieces, toasted

1. Place beets in slow cooker and toss with 1 tablespoon of olive oil; add salt and freshly ground pepper. Cover and cook on high setting until the beets are tender but still hold their shape, about 3 to 3$\frac{1}{2}$ hours.

2. Allow beets to cool, then cut into bite-size pieces. Toss beets with shallots and 1 tablespoon of olive oil; set aside.

3. Combine horseradish, mustard, cream, 1 teaspoon of lemon juice, and 1 tablespoon of olive oil. Season to taste with salt and pepper. Whisk together until well blended.

4. Combine remaining 2 teaspoons of lemon juice and 1 tablespoon of olive oil; toss with lettuce in a large bowl.

5. Arrange beets over lettuce. Drizzle with the horseradish cream. Sprinkle with walnuts.

Makes 6 servings.
Preparation time: 10 minutes
Cooking time: 3 to 3$\frac{1}{2}$ hours
Slow cooker size: Medium; round

Roasted Spring Vegetables

2 slices bacon

5 cloves garlic, minced

1 pound fresh mushrooms, sliced

3 pounds new red potatoes, halved

$\frac{1}{2}$ teaspoon salt

2 cups water

1 tablespoon melted butter

$\frac{1}{2}$ teaspoon salt

$\frac{1}{2}$ teaspoon pepper

$\frac{1}{2}$ teaspoon dill weed

$\frac{1}{2}$ cup water

1 package (16 ounces) frozen peas

12 to 18 spears asparagus, trimmed and cut into 2-inch pieces

✦✦✦

1. Fry bacon in large skillet over medium heat until crisp; remove bacon, leaving drippings. Add garlic and mushrooms; sauté, stirring frequently until golden.

2. Meanwhile, rinse halved potatoes in a mixture of $\frac{1}{2}$ teaspoon of salt and 2 cups of water. Drain and place in slow cooker. Drizzle with melted butter, and toss to coat evenly.

3. Add mushrooms and garlic, then season with $\frac{1}{2}$ teaspoon of salt, pepper, and dill weed. Add $\frac{1}{2}$ cup of water. Crumble bacon and sprinkle over vegetables. Cover and cook on low setting for 6 to 8 hours.

4. About 1 hour before serving, turn slow cooker to high. Add peas and asparagus. Cover and cook for about 1 hour, or until vegetables are tender.

Makes 6 to 8 servings.
Preparation time: 20 minutes
Cooking time: 6 to 8 hours
Slow cooker size: Medium or large; round or oval

Roasted Tomatoes

12 to 15 Roma tomatoes, sliced about $\frac{1}{4}$ inch thick

2 tablespoons extra virgin olive oil

Salt and pepper, to taste

2 tablespoons extra virgin olive oil

$1\frac{1}{2}$ cups fresh bread crumbs

✦✦✦

1. Layer tomatoes in slow cooker. Drizzle with 2 tablespoons of olive oil, and salt and pepper generously. Cover, and cook on high setting for 1 to $1\frac{1}{2}$ hours.

2. Heat remaining 2 tablespoons of olive oil in medium skillet. Add bread crumbs; sauté until golden brown.

3. Sprinkle bread crumbs on top of tomatoes, and continue to cook on high setting for 30 minutes.

Makes 6 servings.
Preparation time: 10 minutes
Cooking time: $1\frac{1}{2}$ to 2 hours
Slow cooker size: Medium;
   round or oval

Savory Sage Bed Dressing

TIPS

Some enjoy a drier dressing, some prefer it a bit more moist. The exact amount of broth you use will also depend on the type, firmness, or freshness of the bread you choose to use.

There just isn't an easier or tastier way to make dressing for your Thanksgiving feast. A tradition at Kathy's house on Thanksgiving morning is to fill a slow cooker with this dressing.

1 loaf (16 ounces) firm-textured white bread, sliced and toasted
¼ cup butter or margarine
2 cups chopped onion
2 cups chopped celery
2 cups sliced mushrooms
2 tablespoons minced fresh parsley
1 teaspoon poultry seasoning
1 teaspoon rubbed sage
½ teaspoon dried thyme leaves
½ teaspoon salt
¼ teaspoon pepper
2 eggs, lightly beaten
1 to 1 ½ cups chicken broth

✦✦✦

1. Tear bread into bite-size pieces and place in large mixing bowl; set aside.
2. Melt butter in large skillet over medium-high heat. Add onions and celery, and sauté for 2 to 3 minutes. Add mushrooms, and continue to cook until vegetables are crisp-tender.
3. Pour vegetable mixture over bread cubes. Sprinkle with parsley and seasonings, and stir well.
4. Add eggs, and stir to blend. Pour chicken broth over bread, and stir to moisten evenly.
5. Spray slow cooker with nonstick spray or butter lightly. Spoon dressing into slow cooker. Cover and cook on high setting for 1 hour. Reduce to low setting, then cook for 4 to 5 hours.

Makes 8 to 10 servings.
Preparation time: 10 to 15 minutes
Cooking time: 5 to 6 hours
Slow cooker size: Large; round or oval

Scalloped au Gratin Potatoes

> *This is the perfect side dish to take to a gathering or potluck supper. It is rich and creamy and oh, so good.*

TIPS

Slicing the potatoes into the salted water helps to ensure that the potatoes don't darken before cooking.

For a thicker sauce, uncover and turn to high during the last 20 to 30 minutes, before adding cream. This allows any extra liquid to boil away.

½ teaspoon salt

2 cups water

6 to 8 medium potatoes, about 3 pounds, peeled and thinly sliced

½ medium onion, chopped

2 tablespoons all-purpose flour

½ teaspoon salt

½ teaspoon pepper

½ teaspoon dry mustard

1 cup chicken broth

2 tablespoons unsalted butter

⅔ cup fresh bread crumbs

1 tablespoon unsalted butter

¾ cup heavy whipping cream

1 tablespoon Dijon mustard

1 cup shredded sharp cheddar cheese

✦ ✦ ✦

1. Mix together ½ teaspoon of salt and water in a mixing bowl. Place sliced potatoes in salted water and toss gently to coat evenly; drain and return to bowl. Add onion.

2. Combine flour, ½ teaspoon of salt, pepper, and mustard; sprinkle over potatoes, and toss to coat well.

3. Spray slow cooker with nonstick spray coating. Place potatoes and onions in slow cooker. Add broth. Dot with 2 tablespoons of butter.

Makes 6 to 8 servings.
Preparation time: 20 minutes
Cooking time: 5 hours
  20 minutes to 7½ hours
Slow cooker size: Medium; round

4. Cover and cook on high setting for 1 hour, then on low setting for 4 to 6 hours, or until potatoes are tender.

5. Meanwhile, melt remaining butter in skillet over medium heat. Add bread crumbs and cook, stirring frequently, until crisp and golden brown. Remove from heat and allow to cool.

6. When potatoes are tender, mix together cream and Dijon mustard; pour over potatoes. Top with cheddar cheese.

7. Cover and cook for 20 to 30 minutes. Sprinkle with crumbs just before serving.

Southwest Cornbread Casserole

Many Tex-Mex and Mexican restaurants accompany their spicy main dishes with a slightly sweet corn and cornbread pudding or casserole. Now you can serve that restaurant flavor at home. Try this the next time you serve tacos or fajitas.

1 can (14.75 ounces) cream-style corn

1 cup frozen corn, partially thawed

½ cup shredded cheddar cheese

1 can (4 ounces) chopped green chiles, drained

½ teaspoon hot pepper sauce

½ teaspoon chili powder

¼ teaspoon garlic powder

¼ teaspoon cumin

1 egg, lightly beaten

⅓ cup milk

⅔ cup yellow cornmeal

½ teaspoon sugar

½ teaspoon salt

¼ teaspoon baking soda

¼ cup shredded cheddar cheese

✳ ✳ ✳

1. Combine cream-style corn, frozen corn, ½ cup of shredded cheese, green chiles, hot pepper sauce, and seasonings in a mixing bowl.

2. Combine egg and milk. Stir in cornmeal, sugar, salt, and baking powder. Blend cornmeal mixture into egg mixture, stirring just until blended. Stir cornmeal batter into corn.

Makes 6 servings.
Preparation time: 10 minutes
Cooking time: 3½ to
   4½ hours
Slow cooker size: Medium;
   round

3. Spray slow cooker with nonstick spray coating. Pour corn mixture into slow cooker. Cover and cook on high setting for 1 hour.

4. Reduce to low setting and cook for $2\frac{1}{2}$ to $3\frac{1}{2}$ hours, or until softly set in center. Sprinkle with cheese; cover and allow cheese to melt (about 5 minutes).

BEANS, GRAINS, RICE

Apple-Cinnamon Oatmeal

TIPS

Steel-cut oats may also be called Irish or Scotch oats. They are oats that are cut into two or three pieces instead of rolled.

> *Imagine waking up and having hot, cooked oatmeal ready to serve from your slow cooker. Perfect for a cold winter's morning.*

1 cup steel-cut oats

4 cups water

½ cup dried apples

1 teaspoon apple pie spice or cinnamon

Pinch of salt

2 tablespoons maple syrup

3 tablespoons toasted chopped pecans, optional

Milk or cream

✦✦✦

1. Combine all ingredients, except nuts and milk, in slow cooker.
2. Cover and cook on low setting for 7 to 9 hours, or overnight.
3. Stir well and, if desired, add pecans at the end of cooking. Serve with milk or cream.

Makes 2 to 3 servings.
Preparation time: 5 minutes
Cooking time: 7 to 9 hours
Slow cooker size: Medium; round

Brown Rice and Lentil Pilaf

This warm and satisfying casserole could be a side dish for a roast, or served as a meatless main dish.

TIPS

The short-grain brown rice in this recipe ensures that the cooked rice will be chewy but nutty. Of course, you may use long-grain brown rice, if preferred.

2 tablespoons olive oil

1 large onion, chopped

5 cloves garlic, minced

2 cups uncooked short-grain brown rice

2 stalks celery, chopped

3 carrots, chopped

1 cup dried lentils

2 cans (14 ounces) vegetable broth

3 cups water

1 bay leaf

2 teaspoons dried oregano leaves

2 teaspoons dried basil leaves

½ teaspoons pepper

½ teaspoons salt

2 tablespoons fresh minced parsley

✷ ✷ ✷

1. Heat olive oil in large skillet over medium-high heat.
2. Add onion and garlic and cook, stirring frequently, until tender.
3. Stir in brown rice and cook, stirring frequently, until rice is toasted.
4. Spoon rice and onions into slow cooker. Stir in remaining ingredients, except parsley.
5. Cover and cook on low setting for 5 to 6 hours, or until rice is tender and liquid is absorbed. Stir in parsley just before serving.

Makes 8 to 10 servings.
Preparation time: 10 to
   15 minutes
Cooking time: 5 to 6 hours
Slow cooker size: Medium;
   round

Old-Fashioned Baked Beans

½ cup ketchup

¼ cup molasses

¼ cup brown sugar

2 tablespoons dry minced onion

½ teaspoon hot pepper sauce

2 tablespoons prepared mustard

2 cans (21 ounces each) pork and beans, with liquid partially drained off

✦✦✦

1. Spray slow cooker with nonstick spray coating. Combine ingredients in slow cooker, and stir gently to combine.

2. Cover, and cook on high setting for 1 to 2 hours. If there is too much liquid, remove lid during last 30 minutes of cooking.

Makes 10 servings.
Preparation time: 5 minutes
Cooking time: 1 to 2 hours
Slow cooker size: Medium;
   round

Slow Cooker Risotto

TIPS

Parmigiano-Reggiano cheese is a quality, aged Italian Parmesan cheese. Try it! Nothing compares to the flavor of this cheese, freshly grated and added to the risotto. It is well worth the minute it will take to grate your own.

> *Let the slow cooker eliminate the tedious work associated with making risotto! This is so easy, it will become a family favorite.*

$\frac{1}{4}$ cup olive oil

2 shallots, peeled and minced

1$\frac{1}{4}$ cups uncooked Arborio rice

$\frac{1}{2}$ cup dry white wine

3$\frac{1}{2}$ cups chicken stock

Dash of salt

$\frac{2}{3}$ to 1 cup freshly grated Parmesan cheese, preferably Parmigiano-Reggiano

✴✴✴

1. Heat oil in a small sauté pan over medium heat.
2. Add shallots and cook, stirring frequently, until shallots have softened. Place in slow cooker.
3. Add rice and toss well to coat. Stir in wine, chicken stock, and salt.
4. Cover and cook on high setting for 2 hours, or until all the liquid is absorbed. Stir in cheese just before serving.

Makes 6 servings.
Preparation time: 10 minutes
Cooking time: 2 hours
Slow cooker size: Medium or large; round or oval

Slow-Simmered Pinto Beans

Each of us has had the opportunity to work with wonderful mentors during our careers. This recipe was inspired by Roxanne's true Texan mentor and dear friend. Thanks, Pat Pitman Smith, for sharing this recipe and so much more with me!

TIPS

If a thicker texture is desired, remove part of the beans just before serving, mash with a fork, and return to slow cooker.

Traditionally, we serve this with cornbread as a main dish, but it is also common to serve it as a side dish with tacos and enchiladas.

1 pound dry pinto beans

9 cups water

2½ teaspoons salt

1 onion, chopped

3 cloves garlic, minced

1 tablespoon chili powder

1 tablespoon Worcestershire sauce

1 to 2 cups chopped ham pieces

1. Rinse beans with cool water, checking for and removing any debris. Place in slow cooker.

2. Add remaining ingredients. Cover, and cook on low setting for 8 to 9 hours.

Makes 6 to 8 servings.
Preparation time: 10 minutes
Cooking time: 8 to 9 hours
Slow cooker size: Medium or large; round or oval

Southwest Chicken and Lentils

TIPS

To speed preparation, go ahead and chop the chicken breasts when you bring them home from the grocery store. Wrap and freeze the chicken, ready for this recipe. Take the chicken out of the freezer and place in the refrigerator the day before cooking this recipe. It will be thawed and ready to go when you are.

> *The earthy, comforting flavor of lentils complements the Southwestern flavors in this dish.*

1 medium onion, chopped

1 stalk celery, chopped

1 carrot, chopped

5 cloves garlic, minced

1 cup lentils

1 pound boneless, skinless chicken breast halves, chopped

1 cup frozen corn

2 cans (14.5 ounces each) chicken broth

2 teaspoons chili powder

1 teaspoon pepper

½ teaspoon dried basil leaves

½ teaspoon dried oregano leaves

½ teaspoon cumin

2 Roma tomatoes, seeded and diced

✳ ✳ ✳

1. Place all ingredients, except tomatoes, in slow cooker.
2. Cover and cook on low setting for 6 to 7 hours.
3. Garnish with diced tomatoes, just before serving.

Makes 6 to 8 servings.
Preparation time: 10 minutes
Cooking time: 6 to 7 hours
Slow cooker size: Medium or
  large; round or oval

Spinach and Artichoke Rice Casserole

TIPS

Be sure to squeeze the spinach dry before adding it to the rice.

If you like spinach-artichoke dip, you will love this casserole! It would be ideal to take to a potluck gathering.

2 tablespoons olive oil

2 shallots, chopped

½ cup chopped onion

2 cloves garlic, minced

1¼ cups uncooked parboiled rice

1 can (14 ounces) vegetable broth

1¾ cups water

1 package (10 ounces) frozen chopped spinach, thawed and well drained

1 can (14 ounces) artichoke hearts, drained and chopped

½ teaspoon salt

¼ teaspoon pepper

¼ teaspoon hot pepper sauce

¼ cup mayonnaise

½ cup sour cream

½ cup grated Parmesan cheese

2 cups shredded Monterey Jack cheese

✦✦✦

1. Spray slow cooker with nonstick spray coating. Heat olive oil in large skillet over medium-high heat. Add shallots, onion, and garlic; cook, stirring frequently, until vegetables are tender.

2. Stir in rice and cook, stirring frequently, until rice is golden. Spoon into slow cooker.

Makes 8 to 10 servings.
Preparation time: 15 minutes
Cooking time: 4 to 5 hours
Slow cooker size: Medium or
   large; round or oval

3. Stir in broth, water, spinach, artichoke hearts, salt, pepper, and hot pepper sauce. Cover, and cook on low setting for 4 to 5 hours.

4. Just before serving, stir in mayonnaise, sour cream, Parmesan cheese, and 1½ cups of shredded Monterey Jack cheese. Sprinkle remaining Monterey Jack cheese on top. Cover and cook for 5 to 10 minutes, or just until cheese is melted.

Tex-Mex Rice Casserole

TIPS

Parboiled rice is often sold under the brand name of "Uncle Ben's Converted Rice." This particular type of rice will hold up better when slow cooking.

2 tablespoons olive oil

1 medium onion, chopped

2 cloves garlic, minced

½ medium green pepper, chopped

1¼ cups uncooked parboiled rice

1 can (14 ounces) vegetable broth

1 can (14.5 ounces) diced tomatoes

1 teaspoon chili powder

½ teaspoon cumin

½ teaspoon salt

¼ teaspoon pepper

1 cup frozen corn

2 tablespoons lime juice

✦✦✦

1. Heat oil in large skillet over medium-high heat. Add onion, garlic, and green pepper; cook, stirring frequently, until onion is tender.

2. Stir in rice and cook, stirring, until rice is golden. Transfer to slow cooker.

3. Stir in remaining ingredients. Cover, and cook on low setting for 4 to 5 hours.

Makes 8 to 10 servings.
Preparation time: 15 minutes
Cooking time: 4 to 5 hours
Slow cooker size: Medium or
    large; round or oval

BREADS

Banana Bread

How can you tell if bread baked in a slow cooker is done? While the top will not brown, it will be set and should be dry. If touched lightly with your finger, the bread should feel firm and bounce back.

The bananas used in banana bread should be quite ripe. Bananas that are too firm and still green will not provide the rich banana flavor needed in the bread.

> *Banana bread, baked in a slow cooker, is perfect. It is moist and tastes so much better than the kind you bake in the oven. Give it a try.*

3 medium, ripe bananas, mashed

2 eggs

¼ cup vegetable oil

¾ cup sugar

2 cups all-purpose flour

1 teaspoon baking soda

¼ teaspoon salt

½ cup chopped walnuts, toasted

✱ ✱ ✱

1. Line bottom of 9-x-5-inch loaf pan or soufflé dish with parchment paper or wax paper. Grease and flour paper.

2. Stir together bananas, eggs, and oil. Stir in sugar, blending well.

3. Stir together flour, soda, and salt; add to banana mixture and stir just until moistened.

4. Stir in walnuts, and pour into prepared pan. Place in slow cooker. Cover and cook on high setting for 2 to 2½ hours, or until done.

Makes 10 servings.
Preparation time: 15 minutes
Cooking time: 2 to 2½ hours
Slow cooker size: Large;
    round or oval

Cinnamon Bread

This bread is great served cool and wonderful when toasted.

3 cups all-purpose flour

½ cup sugar

1 package active dry yeast

1 teaspoon cinnamon

1 teaspoon salt

1 cup warm milk

¼ cup butter, melted

1 egg

1 teaspoon baking powder

Cinnamon topping:

1 tablespoon sugar

1 tablespoon brown sugar

½ teaspoon cinnamon

✦ ✦ ✦

1. Line bottom of 9-x-5-inch loaf pan or soufflé dish with parchment paper or wax paper. Grease and flour paper.

2. Combine flour, sugar, yeast, 1 teaspoon of cinnamon, and salt. In a separate bowl, whisk together milk, butter, and egg. Combine both ingredient bowls, beating the mixture until smooth. Cover and let the batter rest for 1 hour.

3. Stir baking powder into batter. Spoon the batter into prepared pan. Combine cinnamon topping ingredients and sprinkle over batter.

4. Place in slow cooker. Cover and cook on high setting for 2 to 2½ hours, or until done. Allow to stand for 10 minutes, then turn out of pan. Allow to cool completely before slicing.

Makes 10 servings.
Preparation time: 1¼ hours
Cooking time: 2 to 2½ hours
Slow cooker size: Large;
  round or oval

Cranberry Orange Bread

This bread is delicately flavored with orange and dried cranberries. Don't wait for the holidays to enjoy it!

2 cups all-purpose flour

2 teaspoons baking powder

½ teaspoon salt

6 tablespoons butter, softened

¾ cup sugar

2 eggs

½ cup milk

¼ cup orange juice

2 teaspoons grated orange zest

½ cup dried cranberries

½ cup chopped walnuts, toasted

✴✴✴

1. Line bottom of 9-x-5-inch loaf pan or soufflé dish with parchment paper or wax paper. Grease and flour paper.

2. Combine flour, baking powder, and salt; set aside.

3. Beat butter and sugar together with electric mixer until butter is creamy. Beat in eggs. Beat in milk, orange juice and orange zest. Stir in flour mixture, then stir in cranberries and walnuts.

4. Spoon batter into prepared pan and place in slow cooker. Cover and cook on high setting for 2½ to 3½ hours, or until done. Allow to stand for 10 minutes, then turn out of pan.

Makes 10 to 12 servings.
Preparation time: 10 to
 15 minutes
Cooking time: 2½ to
 3½ hours
Slow cooker size: Large;
 round or oval

Oatmeal Batter Bread

1 package active dry yeast

¼ cup warm water

¾ cup milk

¼ cup butter or margarine

¼ cup sugar

1 teaspoon salt

1 egg

2¼ cups all-purpose flour

½ cup quick-cooking rolled oats

+ + +

1. Line bottom of 9-x-5-inch loaf pan or soufflé dish with parchment paper or wax paper. Grease and flour paper.

2. Dissolve yeast in warm water in a large mixing bowl; set aside. Combine milk and butter, and heat until milk is warm and butter is melted (about 120 to 130°).

3. Add warm milk mixture, sugar, salt, and egg to yeast mixture. Beat with mixer until well combined.

4. Stir together flour and oats; add to yeast mixture. Beat at high speed with electric mixer for 1 minute. Pour into prepared pan.

5. Place in slow cooker. Cover and cook on high setting for 2 to 3 hours.

Makes 10 servings.
Preparation time: 15 minutes
Cooking time: 2 to 3 hours
Slow cooker size: Large;
   round or oval

DESSERTS

Best-Ever Chocolate Fondue

TIPS

Fondue is also great with a flavored liqueur stirred into it. Just before serving, stir 2 to 3 tablespoons of chocolate-, almond-, or coffee-flavored liqueur into the chocolate fondue.

> *Fondue is perfect in a slow cooker. The slow cooker keeps the fondue warm, and you get to enjoy the rich flavor without a worry. So, enjoy!*

1 tablespoon butter

2 large milk chocolate bars (8 ounces each), broken into pieces

30 large marshmallows

⅓ cup milk

1 cup heavy whipping cream

Bananas, strawberries, marshmallows, pound cake cubes, pretzels

✦ ✦ ✦

1. Rub slow cooker with butter. Place milk chocolate bars, marshmallows, milk, and cream in slow cooker.

2. Cover and cook on low setting for 1 to 2 hours, stirring every 30 minutes to blend.

3. Serve warm with bananas, strawberries, marshmallows, pound cake, and pretzels.

Makes 6 to 8 servings.
Preparation time: 5 minutes
Cooking time: 1 to 2 hours
Slow cooker size: Small or medium; round

Bread Pudding with Praline Sauce

1 tablespoon butter

6 to 7 cups white bread cubes or 1 loaf (12 ounces) white, French, or Italian bread, cut into 1- to 1½-inch cubes

⅓ cup raisins

1 can (12 ounces) evaporated milk

¼ cup milk

2 eggs

¾ cup sugar

2 teaspoons vanilla

½ teaspoon cinnamon

Praline sauce (recipe following)

Sweetened whipped cream, optional

✳ ✳ ✳

1. Line bottom of slow cooker with parchment paper. Spread butter evenly over the bottom and up the sides of the slow cooker.

2. Arrange bread evenly over the bottom; sprinkle with raisins.

3. Whisk together evaporated milk, milk, eggs, sugar, vanilla, and cinnamon; pour evenly over bread cubes. Gently press bread down into milk mixture to coat evenly.

4. Cover and cook on low setting for 2½ to 3½ hours, or until set and the edges are golden brown.

5. Spoon warm bread pudding into dessert dishes and top each serving with praline sauce. Garnish with whipped cream, if desired.

TIPS

For best results, choose a firm-textured white, unsliced bread. Day-old bread is an excellent choice.

Cook bread pudding until the eggs are set and the edges are golden. Watch carefully so as not to overcook. If edges are browning too quickly the center is not done, turn the slow cooker off and allow it to stand, covered, for 15 minutes or until done in the center.

Makes 8 servings.
Preparation time: 5 to 10 minutes
Cooking time: 2½ to 3½ hours
Slow cooker size: Medium or large; round or oval

Praline Sauce

1 tablespoon butter

¼ cup chopped pecans, toasted

1 cup brown sugar

⅔ cup heavy whipping cream

½ teaspoon vanilla

✦✦✦

1. Melt butter in small saucepan over low heat. Add pecans and cook, stirring frequently, until pecans are golden brown.

2. Stir in brown sugar and cream. Cook, stirring frequently, until mixture boils gently and sugar is dissolved. Remove from heat and stir in vanilla.

Caramel Fondue

TIPS

If desired, substitute $\frac{1}{2}$ teaspoon of rum extract for vanilla.

If desired, stir in 2 to 3 teaspoons of rum or caramel liqueur just before serving.

$\frac{2}{3}$ cup heavy whipping cream

1 package (14 ounces) caramel candies

$\frac{1}{2}$ cup miniature marshmallows

$\frac{1}{2}$ teaspoon vanilla

Apple slices, bananas, pound cake, marshmallows

✦✦✦

1. Lightly butter crock, or spray with nonstick spray coating. Place cream and caramels in slow cooker.

2. Cover and cook on low setting for 1 to 2 hours, or until candy is melted, stirring occasionally.

3. Stir in marshmallows and vanilla, and continue cooking for 30 minutes, or until melted.

4. Serve warm with apple slices, banana pieces, pound cake cubes, or marshmallows.

Makes 6 to 8 servings.
Preparation time: 5 minutes
Cooking time: $1\frac{1}{2}$ to
$2\frac{1}{2}$ hours
Slow cooker size: Small or
medium; round

Caramel Peach Cobbler

The smell of these peaches, simmering in the cinnamon-flavored syrup, will take you back to a childhood visit to Grandma's house.

TIPS

Substitute three bags (16 ounces each) of frozen sliced peaches for fresh ones. No need to thaw them; just pour frozen peaches into the slow cooker. Proceed as recipe directs, but decrease water to ⅓ cup.

These warm, syrupy peaches are so good you can omit the crust if you want. They will taste great spooned over ice cream or cake, or just served in bowls and topped with a crumbled sugar or gingersnap cookie.

Be sure to use a slow cooker with a removable stoneware or crockery bowl. (Never place the electrical heating base of the slow cooker in the oven.) If your slow cooker has a stationary bowl, spoon the peach mixture into a 3-quart casserole dish. Top with rolls and bake as directed.

Makes 8 servings.
Preparation time: 15 minutes
Cooking time: 3 to 4 hours
Baking time: 25 minutes
Slow cooker size: Medium;
    round

8 to 10 cups sliced, peeled, pitted peaches

2 tablespoons lemon juice

1½ cups sugar

2 tablespoons cornstarch

1 teaspoon cinnamon

⅔ cup water

1 tube (11 ounces) refrigerator caramel rolls

½ cup chopped pecans, toasted

1. Place sliced peaches in slow cooker. Drizzle with lemon juice and toss to coat.

2. Combine sugar, cornstarch, and cinnamon; stir into peaches. Add water. Cover and cook on low setting for 3 to 4 hours.

3. Preheat oven to 375°. Open rolls and set caramel aside. Cut rolls in half and arrange over peaches. Drizzle with caramel, then sprinkle with pecans.

4. Remove crockery from slow cooker base and place it in a hot oven. Bake for 25 minutes. Allow to stand for 5 minutes before serving.

Carrot Cake

TIPS

This cake is great with cream cheese frosting (recipe following).

1 cup all-purpose flour

1 cup sugar

1 teaspoon baking soda

$\frac{1}{2}$ teaspoon baking powder

$\frac{1}{2}$ teaspoon cinnamon

$\frac{1}{4}$ teaspoon salt

1 $\frac{1}{2}$ cups grated carrots

2 eggs, lightly beaten

$\frac{1}{2}$ cup vegetable oil

2 tablespoons milk

1 teaspoon vanilla

$\frac{1}{4}$ cup chopped pecans, toasted

✦ ✦ ✦

1. Line bottom of 9-x-5-inch loaf pan or soufflé dish with parchment paper or wax paper. Grease and flour paper.

2. Stir together flour, sugar, baking soda, baking powder, cinnamon, and salt. Add carrots, and toss until combined.

3. Stir together eggs, oil, milk, and vanilla. Add egg mixture to flour mixture, stirring until well combined. Stir in pecans. Pour into prepared pan.

4. Place in slow cooker. Cover and cook on high setting for 2$\frac{1}{2}$ to 3$\frac{1}{2}$ hours, or until set. Allow to stand for 10 minutes; invert, remove pan, and cool.

Makes 8 servings.
Preparation time: 10 minutes
Cooking time: 2$\frac{1}{2}$ to
3$\frac{1}{2}$ hours
Slow cooker size: Large;
round or oval

Cream Cheese Frosting

1 package (3 ounces) cream cheese, softened

¼ cup butter, softened

2¼ cups sifted confectioners' sugar

½ teaspoon vanilla

✳✳✳

1. Beat together cream cheese and butter with electric mixer, beating until creamy. Beat in confectioners' sugar, beating until light and creamy. Beat in vanilla.

2. If desired, split cake horizontally and frost between and on top of the layers.

Chocolate Almond Pound Cake

> *Expect compliments every time you serve this dessert.*
> *Just don't tell anyone how quickly you can mix it up.*

1 package (18.25 ounces) double chocolate muffin mix

1 cup water

1 egg

½ teaspoon almond extract

Glaze:

½ cup sifted confectioners' sugar

½ teaspoon almond extract

2 to 3 teaspoons milk

¼ cup sliced almonds, toasted

✴✴✴

1. Line bottom of 9-x-5-inch loaf pan or soufflé dish with parchment paper or wax paper. Grease and flour paper.

2. Combine muffin mix, water, egg, and almond extract, stirring just until moistened. Spoon batter into prepared pan, and place in slow cooker. Cover and cook on high setting for 2 to 3 hours.

3. Allow to stand for 10 minutes, then turn out of pan. Allow to cool completely on wire rack.

4. Combine confectioners' sugar, almond extract, and milk; stir until smooth. Drizzle over cake, then sprinkle with toasted almonds.

Makes 10 to 12 servings.
Preparation time: 5 minutes
Cooking time: 2 to 3 hours
Slow cooker size: Large;
   round or oval

Chocolate Lava

> This is a wonderful, molten, magic dessert. This dessert ends up with a layer of brownie-like cake and a layer of rich chocolate syrup. Like magic, the syrup you pour on top of the cake seeps to the bottom, so that when you break into the dessert with a spoon, the chocolate syrup oozes through the cake like molten lava.

1 tablespoon butter, softened

1 cup all-purpose flour

¾ cup sugar

3 tablespoons unsweetened cocoa

1½ teaspoons baking powder

Dash salt

½ cup milk

2 tablespoons vegetable oil

1 teaspoon vanilla

½ cup sugar

½ cup brown sugar

¼ cup unsweetened cocoa

1½ cups boiling water

Caramel syrup, optional

Sweetened whipped cream

✦✦✦

Makes 8 servings.
Preparation time: 10 to
   15 minutes
Cooking time: 2 to
   2 ½ hours, plus 30 minutes
   standing time
Slow cooker size: Medium or
   large; round or oval

1. Rub crock with 1 tablespoon of softened butter. Combine flour, ¾ cup of sugar, 3 tablespoons of cocoa, baking powder, and salt. Stir in milk, oil, and vanilla; stir until blended. Spoon into buttered crock.

2. Combine ½ cup of sugar, ½ cup of brown sugar, ¼ cup of cocoa, and boiling water. Stir until sugar is dissolved, then pour gently over batter in slow cooker. Cover and cook on high setting for 2 to 2½ hours, or until top is set.

3. Turn slow cooker off and allow to stand, covered, for 30 minutes. If desired, drizzle with caramel syrup.

4. To serve, spoon from the slow cooker into dessert dishes and top, if desired, with whipped cream.

Decadent Layered Chocolate Cheesecake

> *Cheesecakes can easily crack on the surface, especially if allowed to cool too quickly after baking. This time, let the slow cooker gently bake the cheesecake, then allow both the cheesecake and the slow cooker to cool down slowly. The result will be perfect.*

TIPS

For optimum results, be sure all ingredients, including cream cheese, eggs, cream, and sour cream, are at room temperature before mixing the cheesecake.

Cheesecakes are intended to be rich, wonderful desserts, so this is not the time to choose fat-free or reduced-fat sour cream or cream cheese. These products do not bake the same as regular products and will affect the quality of the finished cheesecake.

Makes 8 servings.
Preparation time: 15 to
 20 minutes
Cooking time: 2½ to 3 hours
Slow cooker size: Large;
 round

Crust:

7 to 8 cream-filled chocolate sandwich cookies

1 tablespoon melted butter or margarine

2 tablespoons brown sugar

Filling:

2 ounces semi-sweet chocolate

2 packages (8 ounces each) cream cheese, softened

⅔ cup sugar

2 eggs

3 tablespoons heavy whipping cream

1 tablespoon all-purpose flour

1 teaspoon vanilla

Topping:

¾ cup sour cream, at room temperature

2 tablespoons sugar

½ teaspoon vanilla

Ganache:

½ cup semi-sweet chocolate chips

⅓ cup heavy whipping cream

¼ teaspoon vanilla

1. Crumple a sheet of aluminum foil, about 24 inches long, into a thin strip, then form into a 7-inch ring. Place aluminum foil ring in slow cooker.

2. For crust: Place cookies in work bowl of food processor. Pulse to crumb. Add melted butter and sugar; pulse to combine. Pour crumbs into 7-inch springform pan. Press crumbs evenly across the bottom of the pan and up the sides about $\frac{1}{2}$ inch; set aside.

3. For filling: Place semi-sweet chocolate in microwave-safe bowl. Microwave for 60 to 90 seconds, stirring after each 30 seconds, or until melted. Set aside to cool slightly.

4. Beat together cream cheese and sugar until well combined. Add eggs and beat well. Scrape bowl and continue beating. Beat in cream, flour, and vanilla.

5. Measure out and set aside 1 cup of batter. Stir melted chocolate into remaining batter. Pour chocolate batter into crust. Pour reserved vanilla batter over chocolate.

6. Place cheesecake in slow cooker, resting squarely on aluminum foil ring. Cover and cook on high setting for $2\frac{1}{2}$ to 3 hours. Do not remove cheesecake from slow cooker.

7. For topping: Blend together sour cream, sugar, and vanilla. Gently spoon sour cream mixture over top of cheesecake, and smooth to cover evenly.

8. Turn slow cooker off. Re-cover slow cooker and allow to stand for about 2 hours, or until cooled almost to room temperature.

9. Lift cheesecake out of slow cooker. Run a knife around the edge and carefully loosen outer ring.

10. For ganache: Combine chocolate chips and cream in small, microwave-safe bowl. Microwave on high power (100%) for about 45 to 60 seconds, stirring midway through, or until cream is hot and chocolate has melted. Blend until smooth. Stir in vanilla.

11. Pour warm ganache over cheesecake, and spread to cover evenly. Refrigerate cheesecake for several hours or overnight.

Pineapple Upside-Down Cake

2 tablespoons butter or margarine, melted

½ cup brown sugar

¼ cup chopped pecans, toasted

1 can (8 ounces) pineapple slices in juice

1 package (9 ounces) yellow cake mix

1 egg

✦ ✦ ✦

1. Line bottom of 9-x-5-inch loaf pan or soufflé dish with parchment paper or wax paper. Grease and flour paper.

2. Pour melted butter into prepared pan. Sprinkle brown sugar and pecans evenly over butter.

3. Drain pineapple slices, reserving juice. Arrange pineapple slices over pecans, overlapping to fit. Measure reserved juice; add water, if necessary to measure ½ cup.

4. Combine yellow cake mix, egg, and ½ cup of pineapple juice. Beat at low speed with electric mixer for 30 seconds, then beat at medium speed for 2 minutes. Pour batter over pineapple slices.

5. Place in slow cooker. Cover and cook on high setting for 2 to 2½ hours. Allow to stand for 10 minutes; invert onto serving platter and serve warm.

Makes 8 servings.
Preparation time: 5 to 10 minutes
Cooking time: 2 to 2½ hours
Slow cooker size: Large; round or oval

Rice Pudding Risotto

TIPS

Arborio rice is now found in most supermarkets and is an Italian, short-grain rice. It is used for risotto and helps create the creamy texture risotto is known to have.

Make this on one of those cold winter days that we sometimes experience in the Midwest. This will warm your loved ones' hearts.

5 cups milk

1 cup uncooked Arborio rice

$\frac{2}{3}$ cup sugar

1 tablespoon vanilla

$\frac{3}{4}$ teaspoon cinnamon

2 heavy dashes nutmeg

Pinch salt

✳ ✳ ✳

1. Lightly butter sides of slow cooker. Place all ingredients into slow cooker, and stir to blend.
2. Cover and cook on low setting for 4 hours, or until thick and creamy, stirring occasionally. Serve warm or cold.

Makes 6 servings.
Preparation time: 5 minutes
Cooking time: 4 hours
Slow cooker size: Medium or
    large; round or oval

Triple Chocolate Pound Cake

2 ounces semi-sweet chocolate

½ cup butter, softened

1 cup sugar

2 eggs

¼ teaspoon baking soda

½ cup buttermilk

1½ cups all-purpose flour

⅓ cup unsweetened cocoa

Dash salt

1 teaspoon vanilla

2 ounces white chocolate

✳✳✳

1. Line bottom of 9-x-5-inch loaf pan or soufflé dish with parchment paper or wax paper. Grease and flour paper.

2. Place semi-sweet chocolate in small, microwave-safe bowl. Microwave on high power (100%) for 30 seconds.

3. Stir, then microwave for an additional 30 seconds, or until melted. Stir until smooth; set aside to cool slightly.

4. Beat butter and sugar together until creamy. Beat in eggs. Stir baking soda into buttermilk, stirring until dissolved.

5. Combine flour, cocoa, and salt in a small bowl. Alternately stir flour mixture and buttermilk mixture into creamed mixture, beginning and ending with flour mixture. Stir in vanilla.

6. Spoon about half of batter into prepared pan. Drizzle with about half of melted semi-sweet chocolate. Top with remaining batter and remaining melted semi-sweet chocolate.

Makes 10 to 15 servings.
Preparation time: 10 to 15 minutes
Cooking time: 2½ to 3½ hours
Slow cooker size: Large; round or oval

7. Use the tip of a knife to swirl melted chocolate through batter. Place in slow cooker. Cover and bake on high setting for $2\frac{1}{2}$ to $3\frac{1}{2}$ hours.

8. Allow to stand for 10 minutes, then turn out of pan. Allow to cool completely on wire rack.

9. Place white chocolate in small, microwave-safe bowl. Microwave on high power (100%) for 30 seconds. Stir, then microwave for an additional 30 seconds, or until melted. Stir until smooth; drizzle over cooled cake.

PART II

CONTACT GRILLS

INTRODUCTION

Quick Sizzle

Broilers and grills have been around a long time, but the contact grill is a fairly new appliance. Often it is referred to by one name, the George Foreman® Grill, marketed by Salton, but actually many models are offered by many companies. By any name, it is a real kitchen workhorse. In all honesty, we discounted its usefulness at first. Then, slowly, we found ourselves using it again. And again. Suddenly, we were totally enthralled. It works. It is quick. It is versatile. It is easy. What more could you want?

What Is a Contact Grill?

Heated grill plates open and close like a waffle iron, so meat, vegetables, and breads cook on both sides at once.

What Do They Do?

- Sear steaks, chops, burgers, and fillets
- Grill vegetables, fruits, and desserts
- Toast breads and sandwiches

What They Don't Do

Contrary to what some report, they do not cook everything perfectly:

- Cuts should be tender or marinated before cooking.
- Sweet sauces and marinades may brown too quickly for some cuts.
- The grill does not ensure weight loss for the diner—oh, how we wish!

Tips for Success

Think Outside the Box

In this case, the box is that square set of grid plates once thought of only as a way to substitute for the outdoor charcoal grill to cook burgers. Wow! You are in for a surprise. The food tastes great and you will enjoy new, wonderful flavors.

Thickness of Food

Select food of the same thickness for even cooking. Don't even try to cook a thick slice and a thin slice at the same time. The top grid must rest on the food evenly.

Removable Grids? Not So Fast!

"Make the grids removable for easier cleanup" is one of those home economics theories that sound good on paper. Yet, it just isn't always true. Removable grids may result in cooler cooking surfaces, and that means less browning, more steaming, and longer cooking times. Grease from the dishwater may also adhere to the bottom of a removable grid, so it smokes or smells the next time it is heated. We generally prefer stationary grids.

Is It Done Yet?

Cooking times for grills will vary, depending on the brand, construction, and size, so cooking times listed here are guides. Be sure to watch the browning carefully, and cook until food is done. If in doubt, use a meat thermometer to check for doneness. Insert an instant-read thermometer into the center or thickest part of the meat. Temperatures should be as follows:

Ground beef or pork	160°
Ground turkey or chicken	165°
Beef	
Medium rare	145°
Medium	160°
Well done	170°
Pork	
Medium	160°
Well done	170°
Poultry	
Chicken breasts, boneless	170°
Chicken thighs, boneless	180°
Turkey breast, boneless	170°

Sizes and Shapes

Cook a single burger patty, or enough for the team. Choose the grill size that fits your family—often one that will cook about four burger patties or chicken breast halves at one time.

Grills—Our Best Advice

Quick and easy is the name of the game, but how can you make the most of this appliance?

■ **Sweet or Savory**

Sweeter sauces and marinades will brown more quickly than those made of vinegar, oil, and such. Some sweeter sauces are actually brushed on during the last moments of cooking.

■ **Toasting Breads**

Buns, breads, and tortillas are easily toasted in the contact grill. You may find that you grill meat, assemble a sandwich or quesadilla, and then want to go right ahead and toast the sandwich. For the safest approach, take the meat off of the grill, turn the grill off, and allow it to cool. When the grids are cool enough, use a heavy towel to wipe off fat or residue. Then, close and preheat the grill again. This way, the bread or tortilla will be crisp, hot, and free from excess fat.

■ **Keep It Closed**

A contact grill is designed to cook on both sides at once. Always place the meat, vegetables, or whatever you are cooking inside the grill, and close the top.

■ **Pound the Chicken**

Chicken breasts are a natural choice for the contact grill. However, for best results, pound them until they are even and about ½ inch thick. Irregular chicken breasts take longer to cook and will cook unevenly.

■ **Bones**

Bones in meat cuts will hold the top of the contact grill up and off of the meat. For best results, use boneless meats.

■ **Single Layers**

Sometimes, with foods such as shrimp or asparagus, one might be tempted to quickly scatter the food onto the grill. It will work so much better, and go faster in the long run, if you use tongs to arrange the food in an even, single layer.

■ **Heat It Up**

When you're in a hurry, does the thought ever occur to you to just skip the preheating? You will regret it. The hot surfaces sear the food, adding to the wonderful flavor and appearance. Plus, cooking will actually take longer without this step. Just be sure to plug the unit in and turn it on as you enter the kitchen and begin to prepare the food. It will be hot when the food is ready to grill.

■ **Cooking in Batches**

Don't crowd the food in a contact grill. In fact, overfilling the contact grill may cause the food to steam more and brown less. It is better to cook food in batches.

■ **It's Hot**

A contact grill cooks quickly. Be careful, as the surfaces, steam, and food will be quite hot. Also, always be sure the drip pan is correctly in place before you start cooking.

APPETIZERS

Chicken Spinach Quesadilla

This recipe is great as an appetizer, but it also makes a quick, easy main dish.

TIPS

Squeeze spinach very dry. To do so, drain spinach in colander. Then, place spinach between several sheets of paper towels and squeeze dry.

If desired, brush tortillas lightly with melted butter instead of spraying with nonstick spray coating.

½ pound boneless, skinless chicken breasts

Salt and pepper, to taste

¼ teaspoon garlic powder

¼ teaspoon ground cumin

1 package (10 ounces) frozen chopped spinach, thawed and drained

1 clove garlic, minced

1 green onion, sliced

2 tablespoons grated Parmesan cheese

½ cup shredded Monterey Jack cheese

1 tablespoon minced fresh cilantro

¼ teaspoon ground cumin

¼ teaspoon salt

8 flour tortillas, about 8 inches in diameter

Shredded Monterey Jack cheese

Minced cilantro

✻ ✻ ✻

1. Preheat contact grill. Pound chicken to ½-inch thickness. Sprinkle chicken with salt, pepper, garlic powder, and ¼ teaspoon of ground cumin.

2. Grill chicken for 5 to 6 minutes, or until meat thermometer registers 170°; allow chicken to cool slightly. Allow grill to cool slightly, wipe clean, and preheat again.

3. Chop chicken. Add spinach, 1 clove of minced garlic, green onion, Parmesan cheese, ½ cup of shredded Monterey Jack cheese, 1 tablespoon of minced cilantro, ¼ teaspoon of cumin, and salt; mix well.

Makes 4 appetizer servings.
Preparation time: 5 minutes
Grill time: 9 to 14 minutes

4. Spray 1 flour tortilla with nonstick spray coating. Place, sprayed-side down, on contact grill. Top with about $\frac{2}{3}$ cup of chicken-spinach mixture.

5. Spray another tortilla with nonstick spray coating and place, sprayed-side up, over filling. Grill for 1 to 2 minutes, or until golden and crisp.

6. Remove to serving plate and immediately sprinkle with shredded Monterey Jack cheese, and garnish with minced cilantro. Cut into quarters. Repeat with remaining filling and tortillas.

Goat Cheese on Toasted Baguettes

TIPS

Rolling the goat cheese in the herbs can be done a day ahead, or even earlier in the day; then just refrigerate the cheese. Allow cheese to warm to room temperature before serving.

2 tablespoons finely chopped chives

2 tablespoons finely chopped Italian parsley

1 log (about 11 ounces) creamy goat cheese

1 baguette, thinly sliced

Olive oil

1. Combine the herbs and place them in a shallow dish. Form goat cheese into a log; roll in fresh herbs.

2. Brush baguette slices with olive oil. Preheat contact grill. Grill bread slices until toasted, and serve with goat cheese.

Makes 8 to 10 appetizer
  servings.
Preparation time: 3 to
  5 minutes
Grill time: 1 to 3 minutes

Grilled Blue Minis

½ cup finely chopped red onion

2 tablespoons butter, softened

2 tablespoons mayonnaise

Salt and pepper, to taste

½ cup crumbled blue cheese

16 slices party rye bread

2 tablespoons melted butter

✦ ✦ ✦

1. Mix together onion, 2 tablespoons of softened butter, mayonnaise, salt, pepper, and blue cheese.

2. Spread on one side of 8 slices of party rye bread. Top with second piece of bread. Brush outside of sandwiches with melted butter.

3. Preheat contact grill. Grill for 1 to 2 minutes, or until bread is toasted and golden.

Makes 8 appetizer servings.
Preparation time: 5 minutes
Grill time: 1 to 2 minutes

Grilled Onion and Blue Cheese Pizzas

TIPS

To toast walnuts, preheat oven to 350°. Spread walnuts in single layer on baking sheet. Bake for 5 to 7 minutes, or until golden.

2 tablespoons olive oil

1 large red onion, cut into ½-inch slices

Salt and pepper, to taste

4 prebaked (7-inch) pizza crusts, Italian bread shells, or small focaccia

2 tablespoons olive oil

2 cloves garlic, minced

2 tablespoons minced fresh rosemary

4 ounces crumbled blue cheese

½ cup chopped walnuts, toasted

✦✦✦

1. Preheat contact grill. Brush 2 tablespoons of olive oil over each side of onion. Grill onion slices for 10 minutes, or until crisp-tender. Remove from grill and place in bowl; season with salt and pepper. Allow to cool slightly.

2. Preheat oven to 425°. Place pizza crusts on baking sheets. Combine remaining 2 tablespoons of olive oil and garlic; brush over top of pizza crusts. Sprinkle with minced rosemary.

3. Separate grilled onions into rings, and arrange over pizza crusts. Top with blue cheese, and sprinkle with walnuts. Bake for 5 to 8 minutes, or until cheese is melted. Cut into wedges and serve.

Makes 16 appetizer servings.
Preparation time: 10 minutes
Grill time: 10 minutes
Baking time: 5 to 8 minutes

Hot Chicken Strips

1 to 1½ pounds boneless, skinless chicken breast halves

2 tablespoons vegetable oil

¼ teaspoon crushed red pepper flakes

Salt and pepper, to taste

½ cup ketchup

¼ cup balsamic vinegar

2 tablespoons brown sugar

2 tablespoons Worcestershire sauce

1 tablespoon Dijon mustard

4 cloves garlic, minced

2 teaspoons paprika

2 teaspoons chili powder

1 teaspoon hot pepper sauce

½ teaspoon salt

½ teaspoon prepared horseradish, optional

1. Pound chicken to ½-inch thickness, then cut into strips about 1 inch wide.

2. Place chicken strips in zip-top bag. Drizzle chicken with oil, and sprinkle with red pepper flakes. Sprinkle with salt and pepper. Seal, and allow to stand for 15 minutes.

3. Meanwhile, combine remaining ingredients in small saucepan. Cook for 5 to 10 minutes, or until hot, stirring frequently.

4. Preheat contact grill. Grill chicken strips for 5 to 6 minutes, or until meat thermometer registers 170°. Serve chicken strips warm, on toothpicks, and accompany with warm sauce.

Makes 8 to 10 appetizer servings.
Preparation time: 15 minutes
Grill time: 5 to 6 minutes

Spicy Steak Strips

These beef strips are wonderful—with just enough spice and flavor to become habit forming. Be sure and serve them with the dipping sauce for a great appetizer.

TIPS

While strip steaks may be a more expensive cut than some families will often choose to serve, serving them as an appetizer is a great way to stretch two steaks to serve several people.

2 beef strip steaks, about 8 ounces each, cut 1¼ inches thick

½ cup beer

1 teaspoon paprika

½ teaspoon pepper

½ teaspoon dry mustard

½ teaspoon garlic powder

½ teaspoon salt

½ teaspoon sugar

¼ teaspoon cayenne pepper

Sweet Mahogany Sauce (see recipe, page 304)

1. Place beef in zip-top bag. Add beer; seal, refrigerate, and allow to marinate for several hours or overnight.

2. Preheat contact grill. Drain and discard marinade. Combine remaining ingredients, except dipping sauce. Sprinkle about one-fourth of seasoning mixture over each side of steak, and rub in gently.

3. Grill steak for 6 to 8 minutes, or until meat thermometer registers 160° for medium doneness. Cover and allow steak to stand for 5 to 10 minutes.

4. Thinly slice steak. Serve warm, on toothpicks. Accompany steak strips with Sweet Mahogany Sauce.

Makes 8 appetizer servings.
Preparation time: 5 minutes
Grill time: 6 to 8 minutes

Tomato and Olive Bruschetta

TIPS

An easy way to peel tomatoes is to first lightly slash an "X" on the bottom of each. Place in boiling water for 5 to 10 seconds, then transfer to ice water and allow to stand for 1 minute. The skins can be pulled off easily.

3 ripe tomatoes, peeled, seeded, and chopped

¼ cup extra virgin olive oil

3 tablespoons pitted, chopped kalamata olives or green olives

3 tablespoons chopped fresh basil

2 cloves garlic, minced

Splash of red wine vinegar

Salt and freshly ground pepper to taste

12 slices Italian bread, about ½ inch thick

Additional olive oil to brush on bread

✦ ✦ ✦

1. Combine tomatoes, ¼ cup of olive oil, olives, basil, garlic, vinegar, salt, and pepper. Toss gently and set aside.

2. Brush both sides of bread lightly with olive oil. Preheat contact grill. Grill bread until toasted; serve tomato mixture on top of bread slices.

Makes 4 to 6 appetizer
servings.
Preparation time: 5 minutes
Grill time: 1 to 3 minutes

SANDWICHES

Bistro Beef and Olive Sandwiches

TIPS

Try new breads for new sandwich flavors. Bakeries often have great olive, cheese, or sun-dried tomato breads that would be great for this sandwich.

⅓ cup chopped, pitted ripe olives

2 tablespoons chopped green onions

2 tablespoons olive oil vinaigrette salad dressing

1 teaspoon dried basil leaves

8 slices firm-textured white bread

2 teaspoons Dijon mustard

8 to 12 thin slices deli-style roast beef

½ cup coarsely chopped, roasted red pepper, drained

2 tomatoes, sliced

4 thin slices provolone cheese

Butter or margarine, softened

✦ ✦ ✦

1. Combine olives, green onions, dressing, and basil; set aside.

2. Lightly spread 4 slices of bread with Dijon mustard. Top each with 2 to 3 slices of meat.

3. Divide olive mixture and roasted red peppers evenly over the sandwiches. Top with tomatoes, cheese, and remaining bread. Lightly spread outside of sandwich with softened butter.

4. Preheat contact grill. Grill sandwiches for 2 to 3 minutes, or until toasted and golden.

Makes 4 servings.
Preparation time: 5 minutes
Grill time: 2 to 3 minutes

Cajun-Grilled Shrimp PO' Boys

> *Warm and spicy grilled shrimp on buns makes a great dinner. Add the flavored mayonnaise spread, and you will be tempted to lick your fingers!*

TIPS

If a spicier sandwich is desired, add ½ teaspoon of hot pepper sauce to shrimp marinade.

1½ pounds large shrimp, peeled and deveined

2 tablespoons olive oil

2 teaspoons Cajun seasoning

½ cup mayonnaise

1 clove garlic, minced

1 green onion, chopped

1 tablespoon Dijon mustard

1 teaspoon Cajun seasoning

2 to 3 drops hot pepper sauce, or to taste optional

2 tablespoons minced fresh parsley

6 sub sandwich buns, toasted

6 lettuce leaves

✦✦✦

1. Place shrimp in zip-top bag; drizzle with olive oil, and sprinkle with 2 teaspoons of Cajun seasoning. Seal bag, and shake to coat evenly. Refrigerate for 20 to 30 minutes.

2. Preheat contact grill. Mix together mayonnaise, garlic, green onion, mustard, 1 teaspoon of Cajun seasoning, hot pepper sauce, and parsley; set aside.

3. Drain shrimp, and discard marinade. Arrange shrimp in single layer on grill; grill for 2 to 3 minutes, or until shrimp turn pink. (If necessary, cook shrimp in batches to avoid overcrowding.)

4. Spread mayonnaise mixture over each bun. Top each with lettuce leaf and shrimp.

Makes 6 servings.
Preparation time: 20 minutes
Grill time: 2 to 3 minutes

California Grilled Cheese Sandwiches

2 cups shredded Monterey Jack cheese

¼ cup pickled jalapenos, drained and finely chopped

8 slices sourdough bread

4 slices deli-style, fully cooked turkey

4 strips crisp bacon

1 avocado, thinly sliced

✦✦✦

1. Combine cheese and jalapenos. Spread one side of each slice of bread with cheese mixture.

2. Place turkey, bacon, and avocado on top of cheese mixture on 4 slices of bread. Top with remaining bread slices, cheese-spread-side down.

3. Preheat contact grill. Grill for 3 to 5 minutes, or until hot and cheese melts.

Makes 4 servings.
Preparation time: 5 minutes
Grill time: 3 to 5 minutes

Chicken Caesar Pitas

2 boneless, skinless chicken breast halves

¼ cup Italian salad dressing

2 tablespoons olive oil

1 clove garlic, minced

4 pita flat breads

6 cups torn romaine

⅓ cup Caesar salad dressing

⅓ cup shredded Parmesan cheese

Freshly ground black pepper

2 Roma tomatoes, seeded and chopped

✦✦✦

1. Pound chicken to ½-inch thickness. Place chicken in zip-top bag, and add Italian salad dressing. Seal, turn to coat well, and refrigerate for 30 minutes.

2. Preheat contact grill. Mix together olive oil and garlic; brush oil evenly on both sides of pita. Grill pitas for about 1 minute, or until hot. Place on plate, cover with kitchen towel, and keep warm.

3. Drain chicken, and discard marinade. Grill chicken for 5 to 6 minutes, or until meat thermometer registers 170°. Set chicken aside, and allow to stand for 5 minutes.

4. Toss together romaine and Caesar salad dressing. Add Parmesan cheese and pepper, and toss well. Place a pita bread on each serving plate. Top each with romaine mixture. Slice chicken into thin strips, and arrange over romaine. Top with chopped tomato.

Makes 4 servings.
Preparation time: 5 to 10 minutes
Grill time: 9 to 10 minutes

Chicken Fajita Wraps

TIPS

Add sliced, pitted ripe olives, sliced green onions, or a dollop of guacamole to each wrap, if desired.

2 boneless, skinless chicken breast halves

⅓ cup freshly squeezed lime juice

⅓ cup olive oil

2 cloves garlic, minced

½ teaspoon cumin

1 onion, sliced ½ inch thick

1 red or green pepper, sliced ½ inch thick

2 tablespoons olive oil

Salt and pepper, to taste

2 tablespoons minced fresh cilantro

¼ cup softened, whipped cream cheese

4 flour tortillas, about 8 to 10 inches in diameter

4 slices Pepper Jack cheese

✦ ✦ ✦

1. Pound chicken to ½-inch thickness. Place chicken in zip-top bag. Combine lime juice, ⅓ cup of olive oil, garlic, and cumin, pour over chicken; seal bag, and refrigerate for 30 minutes.

2. Preheat grill. Drain chicken, and discard marinade. Grill chicken for 5 to 6 minutes, or until meat thermometer registers 170°. Set chicken aside.

3. Brush onion and pepper slices with 2 tablespoons of olive oil. Grill vegetables for 5 to 8 minutes, or until crisp-tender. Place vegetables in bowl, and season with salt and pepper; cover, and allow to stand for 2 or 3 minutes to steam.

4. Mix cilantro into cream cheese. Spread cream cheese mixture on one side of each tortilla.

5. Thinly slice chicken, and arrange over each tortilla. Separate onions into rings, and arrange onions and peppers over chicken. Top each with cheese. Fold tortilla over filling.

Makes 4 servings.
Preparation time: 10 minutes
Grill time: 10 to 13 minutes

Cuban Sandwiches

Once popular in Miami, these sandwiches are now popular in sandwich shops throughout the country. The key to making them is the contact grill, so the meat, cheese, and pickles become hot and steamy while the bread crisps.

8 slices firm-textured white bread

Mustard

8 thin slices deli-style, fully cooked ham

8 thin slices deli-style, fully cooked roast pork

8 thin dill pickle slices

8 thin slices provolone or Swiss cheese

$\frac{1}{4}$ cup butter, softened

✦✦✦

1. Spread one side of each slice of bread with mustard. Arrange ham and pork over 4 slices of bread. Top meat with pickles and cheese. Place second slice of bread, mustard-side down, over cheese. Lightly butter outside of sandwich.

2. Preheat contact grill. Grill sandwiches for 2 to 3 minutes, or until golden brown.

Makes 4 servings.
Preparation time: 5 minutes
Grill time: 2 to 3 minutes

Fontina Cheese Sandwiches

> *This is a fun alternative to traditional grilled cheese. Of course, you can substitute your family's favorite cheese for the Italian fontina.*

4 slices rustic country bread

Butter, slightly softened

4 ounces fontina cheese, thinly sliced

¼ cup roasted red peppers, drained

Salt and pepper, to taste

1. Brush one side of each slice of bread with butter. Layer one half of the cheese and about 2 tablespoons of roasted red peppers on unbuttered side of 1 slice of bread. Salt and pepper lightly.

2. Top with bread, unbuttered-side down. Repeat with remaining bread.

3. Preheat contact grill. Grill sandwiches for 3 to 5 minutes, or until bread is golden. Allow sandwiches to cool slightly, and cut on the diagonal.

Makes 2 servings.
Preparation time: 3 minutes
Grill time: 3 to 5 minutes

Grilled Greek Chicken Pitas

No need to go to a deli to pick up great grilled chicken and a distinctive, Greek-style olive and cucumber salad.

1 pound boneless, skinless chicken breast halves

½ cup Italian salad dressing

1 medium tomato, seeded and chopped

1 small cucumber, sliced

1 small red onion, sliced

8 to 10 kalamata olives, pitted and sliced

¼ cup olive oil

1 tablespoon lemon juice

2 tablespoons red wine vinegar

1 clove garlic, minced

Salt and pepper, to taste

4 pita pocket breads, halved

8 small lettuce leaves

½ cup crumbled feta cheese

✱ ✱ ✱

1. Pound chicken to ½-inch thickness. Place chicken in zip-top plastic bag and add dressing. Seal, and refrigerate for 30 minutes.

2. Meanwhile, combine tomato, cucumber, onion, and olives. Whisk together olive oil, lemon juice, vinegar, garlic, salt, and pepper; pour over tomato mixture.

3. Preheat contact grill. Drain chicken, and discard marinade. Grill chicken for 5 to 6 minutes, or until meat thermometer registers 170°. Remove from grill and allow to stand for 5 minutes. Thinly slice chicken into strips.

4. Place a lettuce leaf in each pita pocket. Divide chicken, and spoon into each pocket; top with tomato mixture. Sprinkle with feta.

Makes 4 servings.
Preparation time: 5 to
10 minutes
Grill time: 5 to 6 minutes

Grilled Italian Beef Sandwiches

4 slices sourdough or Italian bread, sliced about ½ inch thick

Softened butter or margarine

2 tablespoons mayonnaise

¼ teaspoon dried basil leaves

4 thin slices Italian-seasoned, deli-style, fully cooked beef, or 4 thin slices Italian beef (see page 102)

2 thin slices tomato

2 thin slices red onions

2 slices provolone cheese

✦✦✦

1. Spread one side of each bread slice with softened butter or margarine. Mix together mayonnaise and basil; spread over other side of bread.

2. Top mayonnaise side of 2 bread slices with meat, tomatoes, onions, and cheese. Top with second slice of bread, mayonnaise-side down.

3. Preheat contact grill. Grill sandwiches for 2 to 3 minutes, or until golden brown.

Makes 2 servings.
Preparation time: 5 minutes
Grill time: 2 to 3 minutes

Honey-Mustard Chicken Sandwiches with Grilled Red Onion Marmalade

TIPS

You can make grilled red onion marmalade a day or two in advance, if desired. Cover and refrigerate until serving time. Allow marmalade to warm to room temperature before serving.

Red onion marmalade is also excellent as a condiment with any grilled meat.

1 medium red onion, sliced $\frac{1}{2}$ inch thick

Olive oil

2 tablespoons red wine vinegar

2 tablespoons orange juice

2 teaspoons sugar

$\frac{1}{2}$ teaspoon salt

4 boneless, skinless chicken breast halves

3 tablespoons mustard

2 tablespoons honey

4 hamburger buns, toasted

✦ ✦ ✦

1. Preheat contact grill. Brush onion slices with olive oil. Grill onion for 10 minutes, or until crisp-tender. Remove onion from grill, and coarsely chop.

2. Place chopped onion in small saucepan, and add red wine vinegar, orange juice, sugar, and salt. Cook for 15 to 20 minutes, or until thick and most of moisture has evaporated, stirring frequently.

3. Meanwhile, pound chicken to $\frac{1}{2}$-inch thickness. Grill chicken for 5 to 6 minutes, or until meat thermometer registers 170°.

4. Combine mustard and honey. Place chicken on buns, and top each with honey mustard and a spoonful of grilled red onion marmalade.

Makes 4 servings.
Preparation time: 20 to
   30 minutes
Grill time: 15 to 16 minutes

Panini Explored

Panini, in the purest sense, is Italian for roll or biscuit, but in today's sandwich culture, it means grilled and fantastic. Restaurants serve all kinds of hot, crisp sandwiches. Now, with your contact grill, all of these taste sensations—once reserved for eating out—are just minutes away from being served at your own dinner table.

Mix, match, and create! Once, ham and cheese meant a shaved deli ham paired with American cheese on white bread. Today, that description may mean prosciutto and Gruyère on artisan bread. More importantly, the list of possible combinations is nearly endless.

Here are some tips to make "Perfect Paninis":

- Assemble the ingredients while the contact grill preheats.
- Grill onions, peppers, chicken, or any other ingredients that are cooked first. Allow the contact grill to cool; wipe clean, preheat again, then grill the sandwich. This only takes minutes, but the result is perfect.
- Assemble the sandwiches, and grill for about 2 to 3 minutes, or until bread is toasted and crisp.

The Panini Fixings

Breads

Artisan types and bakery breads, such as sourdough, ciabatta, honey wheat, or focaccia, make wonderful sandwiches. The flavored breads, including Asiago cheese, sun-dried tomato, olive, onion, pesto, and so many others, make the choices nearly endless. Slice bread about $\frac{1}{2}$ to 1 inch thick. Butter the outside of the bread, if desired. (If using focaccia or other larger loaves of bread, check the fit on a cold grill, then cut to fit, if necessary, before splitting and filling the sandwich.)

Meats

Fully cooked, grilled chicken and turkey are great. Slice pieces of roast beef or pork from those you cook in the slow cooker one night, and keep them for panini the next day. Try new flavors and discover a favorite. All meats should be fully cooked before assembling the sandwich. When choosing deli-sliced meats, several thin slices will taste better than one larger piece, and those that are freshly sliced will have the best flavor.

Cheese

Again, choose thin slices for the best flavor and even melting. Note that those that are softer will melt more quickly than those that are harder or drier. Explore the gourmet cheese counter for new ideas. You only need a few slices, so have the attendant freshly slice a few for you to try. You won't believe the difference, for example, when you try fresh mozzarella as opposed to the pre-shredded, packaged version you have used before. Herb-seasoned cream cheese spreads, goat cheese, or specialty cheeses may also add some new flavors. Even typical grocery stores will often sell Mexican cheeses or other new varieties that will be fun to try.

Vegetables

Grilled or freshly sliced, vegetables make a wonderful flavor contribution. Onions (yellow, white, or red) are naturals for sandwiches. Add peppers (red, green, or yellow), mushrooms, tomatoes, pepperoncini, olives, marinated artichoke hearts, fresh basil leaves, spinach, lettuce, eggplant, zucchini, or even sliced apples or pears. To grill them, slice about $\frac{1}{2}$ inch thick, brush with olive oil, and grill until crisp-tender.

Herbs and Such

Add a sprinkle of fresh minced or dried herbs. A few cranks from a pepper mill will add zing and flavor. Garlic, in any form, is always a favorite. Drizzle the sandwich fillings with balsamic or another vinegar.

Spreads and More

Check out the flavored mayonnaise recipes that precede the panini recipes. Or, forget the mayonnaise, and look at various spreads and flavored oils or vinegars. For example, add a light spread of basil pesto (see recipe, p. 10), olive tapenade, minced, oil-packed sun-dried tomatoes, butter seasoned with herbs such as basil or dill, pizza sauce, or softened cream cheese.

For a flavored mayonnaise, always use the best quality, freshest mayonnaise available. It is worth it to buy a small, fresh jar and discard the one that may have been open too long in the back of the refrigerator. Many recipes will work with a low-fat mayonnaise, but sometimes the flavors do not blend as well with the fat-free varieties. Once mixed, store tightly covered in the refrigerator and use within about 5 to 7 days. Keep experimenting with your own favorites. The ideas are endless, but try some of the following recipes for starters. Then you can move on to blending curry powder, horseradish, minced shallots or green onions, crumbled blue cheese, or grated lemon or lime zest into mayonnaise.

Aioli

This is a classic, garlic-flavored mayonnaise, especially popular in Spanish foods. Use it as a wonderful sandwich spread, or just dollop some next to grilled vegetables, fish, chicken, or beef as a tasty accompaniment.

1 cup mayonnaise

4 cloves garlic, minced

1 tablespoon freshly squeezed lemon juice

Salt and freshly ground black pepper, to taste

1. Mix together all ingredients. Cover, and refrigerate overnight for flavors to blend.

Makes about 1 cup.
Preparation time: 3 minutes

Basil Mayonnaise

This mayonnaise is especially good on chicken or turkey, or served with roasted, grilled vegetable panini.

1 cup fresh basil leaves
½ cup mayonnaise

✳✳✳

1. Place basil leaves in work bowl of food processor. Pulse to chop. Add mayonnaise, and pulse to lightly blend.

Makes about ¾ cup.
Preparation time: 3 minutes

Chipotle Mayonnaise

TIPS:

See note about chipotle peppers on page 96.

1 cup mayonnaise

1 tablespoon minced, canned chipotle pepper in adobo sauce

✦✦✦

1. Combine all ingredients, blending well.

Makes about 1 cup.
Preparation time: 3 minutes

Cream Cheese and Olive Spread

> *Ready for a change of pace? Omit the mayonnaise, and use this tasty spread instead!*

½ cup cream cheese, softened

¼ cup chopped ripe, pitted olives, or pitted kalamata olives

✦ ✦ ✦

1. Beat cream cheese until soft; blend in olives.

Makes about ¾ cup.
Preparation time: 3 minutes

Sour-Cream Mustard Sauce

⅓ cup mayonnaise

⅓ cup sour cream

2 tablespoons Dijon mustard

✦ ✦ ✦

1. Combine all ingredients, blending well.

Makes about ¾ cup.
Preparation time: 3 minutes

Summertime Herb Mayonnaise

TIPS

To quickly and easily mince the herbs, place the parsley and herbs in a straight-sided cup or liquid measuring cup; use kitchen shears, vertically held, to quickly snip the herbs. Generally, begin with about double the amount of loosely packed herb leaves that you want to end up with. For example, in this case, if you want to end up with 2 tablespoons of minced herbs, begin with about $\frac{1}{4}$ cup of fresh herb leaves.

1 cup mayonnaise

2 tablespoons minced fresh parsley

2 tablespoons minced fresh dill, basil, rosemary, oregano, dill, or other fresh herb

✳ ✳ ✳

1. Combine all ingredients, blending well.

Makes about 1$\frac{1}{4}$ cups.
Preparation time: 3 to
   5 minutes

Tomato Mayonnaise

½ cup mayonnaise

2 tablespoons minced, oil-packed sun-dried tomatoes, drained

¼ teaspoon dried basil leaves

✦✦✦

1. Combine all ingredients, blending well.

Makes about ⅔ cup.
Preparation time: 3 minutes

Hot Antipasto Panini

1 can (14.5 ounces) artichoke hearts, rinsed and drained, chopped

1 clove garlic, finely minced

2 tablespoons extra virgin olive oil

½ cup roasted red peppers, drained and chopped

2 tablespoons finely chopped pepperoncini

3 tablespoons finely chopped pitted black olives

Salt and pepper, to taste

1 loaf ciabatta or Italian bread

1 cup arugula

½ pound hard salami or cappicola, sliced

4 slices provolone

✦ ✦ ✦

1. Preheat contact grill. Combine artichoke, garlic, olive oil, red peppers, pepperoncini, and olives. Salt and pepper to taste.

2. Slice bread in half lengthwise; then cut into fourths. Place arugula on bottom half of 4 pieces of bread. Top with salami and cheese.

3. Divide artichoke mixture and place on top of cheese. Top with top piece of bread. Grill for 3 to 5 minutes, or until hot and cheese melts.

Makes 4 servings.
Preparation time: 5 to 10 minutes
Grill time: 3 to 5 minutes

Italian Vegetable Panini

2 slices peeled eggplant, about ½ to ¾ inch thick

Salt

2 tablespoons white wine vinegar

1 tablespoon lemon juice

1 teaspoon olive oil

1 teaspoon Italian seasoning

¼ teaspoon garlic powder

¼ teaspoon salt

1 Roma or small tomato, thinly sliced

½ medium red pepper

4 whole button mushrooms

2 slices red or yellow onion, about ½ inch thick

1 tablespoon mayonnaise

¼ teaspoon garlic powder

¼ teaspoon dried basil leaves

4 slices firm-textured Italian or country-style white bread

2 slices provolone cheese

Butter, softened

✦ ✦ ✦

1. Generously sprinkle eggplant slices with salt; allow to stand for 30 minutes in a colander, then rinse and drain well.

2. Mix together vinegar, lemon juice, olive oil, Italian seasoning, garlic powder, and ¼ teaspoon of salt. Place sliced tomato in a bowl, and drizzle with 1 tablespoon of vinegar mixture; set aside.

3. Slash bottom of red pepper so it will lay flat. Place red pepper, mushrooms, onion slices, and eggplant in a zip-top bag. Drizzle with remaining vinegar mixture; seal, and allow to stand for 10 minutes.

Makes 2 servings.
Preparation time: 40 minutes
Grill time: 13 minutes

(continues on next page)

4. Preheat contact grill. Drain marinade from eggplant, pepper, mushrooms, and onions, then arrange vegetables in contact grill. (Do not grill tomatoes.) Grill for 5 minutes. Remove eggplant and mushrooms from grill; place on plate and keep warm.

5. Continue grilling pepper and onion for an additional 5 minutes, or until tender. (Grill vegetables in batches, if necessary.) Turn contact grill off; cool, and wipe clean.

6. Preheat contact grill again. Mix together mayonnaise with $\frac{1}{4}$ teaspoon of garlic powder and basil. Spread mayonnaise on 1 side of 2 slices of bread. Top each with 1 slice provolone cheese. Cut grilled pepper and mushrooms into thin slices, and arrange over cheese. Top with eggplant slices and onion slices. Drain tomatoes, and arrange over grilled vegetables. Top with remaining bread.

7. Lightly butter outside of sandwich. Grill for about 3 minutes, or until sandwiches are golden brown.

Mediterranean Steak and Artichoke Sandwiches

1 pound boneless beef sirloin, cut about $\frac{1}{2}$ to $\frac{3}{4}$ inch thick

1 jar (6.5 ounces) quartered and marinated artichoke hearts

2 tablespoons red wine vinegar

1 green onion, chopped

1 clove garlic, minced

$\frac{1}{2}$ teaspoon red wine vinegar

$\frac{1}{2}$ teaspoon dried basil leaves

$\frac{1}{4}$ teaspoon salt

$\frac{1}{8}$ teaspoon pepper

2 tablespoons grated Parmesan cheese

$\frac{1}{2}$ cup roasted red peppers, drained and chopped

8 slices Italian bread, cut about $\frac{3}{4}$ inch thick

Butter, softened

1. Place beef in zip-top bag. Drain artichoke hearts, reserving liquid. Blend 2 tablespoons of red wine vinegar into reserved liquid; pour mixture over steak. Seal bag, and refrigerate for several hours.

2. Meanwhile, chop artichoke hearts. Combine chopped artichoke hearts, green onion, garlic, $\frac{1}{2}$ teaspoon of red wine vinegar, basil, salt, pepper, and Parmesan cheese. Cover, and refrigerate until serving time.

3. Preheat contact grill. Drain steak, and discard marinade. Grill steak for about 5 minutes for medium doneness. Remove steak to cutting board; cover, and allow to stand for about 5 to 10 minutes.

4. Turn grill off, and allow to cool. When cool, wipe clean and then preheat again. When steak is cool enough to handle, thinly slice it across the grain.

Makes 4 servings.
Preparation time: 10 minutes
Grill time: 8 minutes

(continues on next page)

5. To assemble sandwiches, top each of 4 slices of bread with $\frac{1}{4}$ of the meat slices, about 2 tablespoons of chopped roasted red pepper, and about $\frac{1}{4}$ of the artichoke relish. Top with remaining slice of bread. Lightly butter outside of sandwich.

6. Grill sandwiches for about 3 minutes, or until browned and crusty.

Open-Face Sausage Kabob Sandwiches

2 tablespoons red wine vinegar

1 clove garlic, minced

1 teaspoon Dijon mustard

Salt and pepper

⅓ cup extra virgin olive oil

2 peppers (one red and one green), seeded and cut into 1½-inch pieces

8 whole button mushrooms

1 medium onion, cut into 1½-inch pieces

1 pound Italian sausage links, cut into 1½-inch pieces

8 slices sourdough bread, cut into slices about 1 inch thick

1 bag (10 ounces) salad greens

1. Whisk together vinegar, garlic, and Dijon mustard. Season to taste with salt and pepper. Whisk in olive oil.

2. Preheat contact grill. Toss peppers, onions, and mushrooms with 3 tablespoons of vinegar mixture. Thread vegetables and sausage onto skewers. Grill for 5 to 6 minutes or until sausage is fully-cooked.

3. Grill bread slices to quickly toast. Drizzle each slice of bread with a small amount of dressing. Toss remaining dressing with greens. Place toasted bread in center of each serving plate. Mound salad around bread, and top with kabob.

Makes 4 servings.
Preparation time: 5 to 10 minutes
Grill time: 6 to 7 minutes

Ranch Chicken and Bacon Wraps

TIPS

The chicken cooked with the peppered bacon is great. If desired, serve on a toasted hamburger bun and top with lettuce, sliced tomatoes, and honey mustard.

2 boneless, skinless chicken breast halves

½ cup sour cream

¼ cup mayonnaise

1 tablespoon minced fresh parsley

¼ teaspoon pepper

¼ teaspoon onion salt

¼ teaspoon garlic powder

4 slices thick-sliced, peppered bacon

1 teaspoon Dijon mustard

4 flour tortillas, about 10 inches in diameter, warmed

1 cup shredded lettuce

½ cup shredded carrot

2 Roma tomatoes, chopped

½ cup shredded Monterey Jack cheese

✤ ✤ ✤

1. Pound chicken breasts to ½-inch thickness, then place in zip-top plastic food bag.

2. Combine sour cream, mayonnaise, parsley, pepper, onion salt, and garlic powder. Spoon about half of mixture over chicken; reserve remaining half. Seal and squeeze bag to cover chicken thoroughly.

3. Preheat contact grill. Cut bacon in half. Arrange 2 half-pieces of bacon closely together on grill. Place a chicken breast on top of bacon, covering bacon completely. Top chicken with 2 more half-pieces of bacon. Repeat with second chicken breast and remaining bacon.

4. Grill for 5 to 6 minutes, or until meat thermometer registers 170° and bacon is crisp. Set aside until just cool enough to handle. Thinly slice chicken and bacon into strips.

5. Blend mustard into reserved sour cream mixture. Spread over one side of each warm tortilla. Top each with lettuce, carrot, tomatoes, and cheese. Add chicken and bacon strips. Roll tortilla over filling.

Makes 4 servings.
Preparation time: 10 minutes
Grill time: 5 to 6 minutes

Spring Turkey Wraps

Turkey will become a year-round favorite with this great wrap sandwich.

1 pound boneless, skinless turkey breast, cut into slices about ½ inch thick

2 tablespoons olive oil

Salt and pepper, to taste

8 stalks asparagus, trimmed

1 tablespoon olive oil

¼ cup softened, whipped cream cheese

4 flour tortillas or spinach-flavored tortillas, about 8 to 10 inches in diameter

6 to 8 fresh spinach leaves

4 slices cheddar cheese

½ cup roasted red peppers, drained

✦✦✦

1. Preheat contact grill. Brush each turkey piece with olive oil, and season with salt and pepper.

2. Grill turkey for 5 to 8 minutes, or until meat thermometer registers 170°. Set turkey aside.

3. Toss asparagus spears in 1 tablespoon of olive oil; grill asparagus for 3 minutes, or until crisp-tender. Season asparagus with salt and pepper.

4. Spread cream cheese over one side of each tortilla. Arrange spinach leaves and cheese over cream cheese.

5. Slice turkey into thin strips, and arrange over cheese. Top with grilled asparagus and roasted red peppers. Wrap tortilla around filling.

Makes 4 servings.
Preparation time: 5 minutes
Grill time: 8 to 11 minutes

Toasted Garlic Cheese Strips

If you like garlic bread, then this recipe is for you. Serve these tasty sandwich strips with pasta, salad, or soup.

TIPS

For a tasty appetizer, serve Toasted Garlic Cheese Strips with warm Marinara Sauce as a dipping sauce. See recipe for Marinara Sauce on page 55.

¼ cup butter, softened

2 cloves garlic, minced

½ teaspoon dried basil leaves

¼ cup grated Parmesan cheese

8 slices Italian bread

4 slices mozzarella cheese

1 egg

½ cup milk

¾ cup Italian-seasoned bread crumbs

✦✦✦

1. Stir together butter, garlic, basil, and Parmesan cheese. Spread butter mixture over 4 slices of Italian bread. Top butter mixture with a slice of mozzarella cheese. Top with second slice of bread.

2. Preheat contact grill. Whisk together egg and milk; pour into pie plate. Place bread crumbs on a plate. Quickly dip sandwich into milk mixture, turning to coat evenly, then coat evenly on both sides with bread crumbs.

3. Grill for 1 to 2 minutes, or until golden. (Grill 2 sandwiches at once, if necessary, to avoid overcrowding.)

4. Allow sandwiches to cool for about 1 minute. With a sharp knife, slice each sandwich in half; then slice each in half again to form 4 strips.

Makes 8 servings.
Preparation time: 5 minutes
Grill time: 1 to 2 minutes

Tuscan Chicken Sandwiches

2 boneless, skinless chicken breast halves

2 tablespoons Italian salad dressing

2 tablespoons minced fresh basil

1 clove garlic, minced

4 tablespoons softened, whipped cream cheese

4 slices Asiago or olive bread

4 thin slices Genoa salami

$\frac{1}{4}$ cup kalamata olives, pitted and sliced

$\frac{1}{4}$ cup roasted red peppers, drained

Softened butter

1. Preheat contact grill. Pound chicken until very thin, about $\frac{1}{4}$ inch thick. Brush Italian salad dressing on each side of chicken.

2. Grill chicken for about 5 minutes, or until meat thermometer registers 170°. Remove chicken from grill and set aside; allow to stand for 5 minutes. Allow contact grill to cool, and wipe clean.

3. Mix basil and garlic into cream cheese. Spread cream cheese mixture on one side of each slice of bread. Top bread with chicken, then top each with salami, olives, and red peppers. Top with remaining bread, cream cheese side down. Spread outside of each sandwich with butter.

4. Grill for 2 minutes, or until sandwich is golden brown.

Makes 2 servings.
Preparation time: 5 minutes
Grill time: 7 minutes

BURGERS

Buffalo Burgers

Buffalo—as in the spicy flavor! Add that famous flavor to beef burgers, and get ready for great eating.

1 pound ground chuck

Salt and pepper, to taste

1 tablespoon melted butter

5 teaspoons hot pepper sauce

4 hamburger buns, split and toasted

⅓ cup sliced celery

¼ cup blue cheese salad dressing

1. Preheat contact grill. Season ground chuck with salt and pepper. Form into 4 patties.
2. Grill for 5 to 8 minutes, or until meat thermometer registers 160°. Combine butter and hot pepper sauce. Before removing burgers from grill, brush burgers generously with hot sauce mixture.
3. Place burgers on buns. If desired, top with remaining hot sauce. Sprinkle with celery, and dollop with salad dressing.

Makes 4 servings.
Preparation time: 3 minutes
Grill time: 5 to 8 minutes

Caesar Hamburgers

1 pound ground chuck

Coarsely ground pepper

Salt, to taste

4 hamburger buns, split and toasted

8 large shavings Parmesan cheese

4 tablespoons Caesar salad dressing

+ + +

1. Preheat contact grill. Add at least 2 teaspoons coarsely ground pepper to ground chuck; salt to taste. Shape into 4 patties.

2. Grill for 5 to 8 minutes, or until meat thermometer registers 160°. Place burger on bun, and top each with Parmesan cheese and 1 tablespoon of Caesar dressing.

Makes 4 servings.
Preparation time: 3 minutes
Grill time: 5 to 8 minutes

Francophile Hamburgers

1 pound ground chuck

¼ cup burgundy wine

½ teaspoon dried thyme leaves

Salt and freshly ground pepper, to taste

4 thick slices French bread, toasted

4 tablespoons garlic cheese spread

✦✦✦

1. Combine ground chuck, wine, thyme, salt, and pepper. Shape into 4 patties.

2. Preheat grill. Grill beef patties for 5 to 8 minutes, or until meat thermometer registers 160°. Place beef patties on top of French bread, and dollop with cheese spread.

Makes 4 servings.
Preparation time: 5 minutes
Grill time: 5 to 8 minutes

Grilled Greek Burgers

TIPS

One key to the success of this recipe is fresh pita. With the freshest pita, the kind that just melt in your mouth, these burgers are scrumptious. If fresh pita breads aren't available, substitute toasted hamburger buns.

Everyone loves burgers, but sometimes they fall into the "same old" category. Not these! Flavored with mint, cumin, and onion, then topped with grilled red onions and yogurt, the flavor shouts "new and exciting"!

1 large red onion, cut into ½-inch slices

Olive oil

1½ cups grape tomatoes, cut in half

1 tablespoon minced fresh mint

Splash cider vinegar

Splash extra virgin olive oil

¼ cup finely chopped onion

¼ cup finely chopped fresh parsley

2 cloves garlic, minced

1½ teaspoons ground cumin

½ teaspoon salt

½ teaspoon coarsely ground black pepper

1¼ pounds ground chuck

Pita flat bread

Yogurt Sauce (recipe following)

✳ ✳ ✳

1. Preheat contact grill. Brush red onion slices with olive oil and grill until tender. Coarsely chop grilled red onion and toss with tomatoes and mint. Splash with vinegar and olive oil; mix well. Cover and set aside.

Makes 4 servings.
Preparation time: 10 to 15 minutes
Grill time: 10 to 13 minutes

(continues on next page)

2. Combine finely chopped onion, parsley, garlic, cumin, salt, pepper, and ground chuck. Form into 4 patties. Grill for 5 to 8 minutes, or until meat thermometer registers 160°.

3. While burgers are cooking, wrap pita bread in aluminum foil and warm in 300° oven for 5 to 10 minutes.

4. Place burger on top of pita and top with tomato mixture. Serve with yogurt sauce. Roll tightly, almost as if preparing a taco using the pita.

Yogurt Sauce

½ cup plain yogurt

1 teaspoon lemon juice

1 teaspoon extra virgin olive oil

1 teaspoon dried dill weed

✦ ✦ ✦

1. Combine all ingredients and refrigerate.

Hamburgers with Caramelized Onions

1 pound ground chuck

½ teaspoon dried thyme leaves

½ teaspoon dried oregano leaves

Salt and pepper, to taste

4 hamburger buns, split and toasted

½ cup caramelized onions (see recipe, page 12)

TIPS

Make sure to heat caramelized onions before adding to hamburger.

✦✦✦

1. Preheat contact grill. Combine ground chuck, thyme, oregano, salt, and pepper. Form into 4 patties.

2. Grill for 5 to 8 minutes, or until meat thermometer registers 160°. Place on hamburger bun; top with caramelized onions.

Makes 4 servings.
Preparation time: 5 minutes
Grill time: 5 to 8 minutes

Italian Burgers with Grilled Portobello Mushroom

1 pound ground beef

½ teaspoon garlic salt

½ teaspoon Italian seasoning

1 portobello mushroom

2 tablespoons olive oil

Toasted hamburger buns

4 thin slices mozzarella or provolone cheese

1 tomato, thinly sliced

✦✦✦

1. Preheat contact grill. Mix together ground beef, garlic salt, and Italian seasoning; shape into 4 patties.

2. Slice portobello mushroom into slices about ½ inch thick; brush both sides with olive oil. Grill mushroom slices for 3 to 5 minutes. Remove from grill; cover to keep warm.

3. Grill patties for 5 to 8 minutes, or until meat thermometer registers 160°. Place on buns and immediately top hot burgers with cheese. Divide mushrooms evenly over burgers, and top each with a thin slice of tomato.

Makes 4 burgers.
Preparation time: 5 minutes
Grill time: 8 to 13 minutes

Meatloaf Burgers

1 pound ground chuck

1 egg

½ cup dry bread crumbs

⅓ cup tomato sauce

½ cup finely chopped onion

Salt and pepper, to taste

4 hamburger buns, split and toasted

4 strips bacon, cooked until crisp

4 tablespoons chili sauce

✦✦✦

1. Preheat contact grill. Combine ground chuck, egg, bread crumbs, tomato sauce, onion, salt, and pepper; blend well. Form into 4 patties.

2. Grill for 5 to 8 minutes, or until meat thermometer registers 160°. Place patties on hamburger buns; top with bacon and chili sauce.

Makes 4 servings.
Preparation time: 5 minutes
Grill time: 5 to 8 minutes

Pepper-Crusted Burgers with Grilled Balsamic Onions

Sometimes adults enjoy burgers that are different from those the kids crave. These burgers, fragrant and tasty, with pepper and grilled onions, just might be the ones to serve when adults gather.

1 pound ground beef

1 teaspoon Worcestershire sauce

Freshly ground pepper

1 large sweet onion, cut into ½-inch slices

Olive oil

1 teaspoon balsamic vinegar

Toasted hamburger buns

1. Preheat contact grill. Mix together ground beef and Worcestershire sauce. Shape into 4 patties. Sprinkle each generously with freshly ground pepper.

2. Brush onion slices with olive oil. Grill onion for 8 to 9 minutes, or until well browned.

3. Remove onion and place in a small bowl. Drizzle with balsamic vinegar, and toss to coat. Cover bowl with aluminum foil, and set aside to allow onions to steam.

4. Cook ground beef patties for 5 to 8 minutes, or until meat thermometer registers 160°. Serve beef patties on buns, and top each with grilled onions.

TIPS

If you have caramelized onions in the freezer (see page 12), this would be a great time to use them instead of the grilled onions in this recipe. Thaw ⅔ cup of onions. To serve, warm onions by heating in microwave oven 1 to 2 minutes, or until hot. Drizzle with balsamic vinegar.

Makes 4 servings.
Preparation time: 5 minutes
Grill time: 13 to 17 minutes

South-of-the-Border Burgers

1 pound ground chuck

½ teaspoon salt

¼ teaspoon pepper

1 teaspoon chili powder

½ teaspoon cumin

4 slices Pepper Jack cheese

4 hamburger buns, split and toasted

1 cup refried beans

Salsa

Guacamole

Pickled, sliced jalapeno peppers

Crushed tortilla chips

✦✦✦

1. Preheat contact grill. Place ground chuck in bowl. Add salt, pepper, chili powder, and cumin; blend well. Form into 4 patties. Grill for 5 to 8 minutes, or until meat thermometer registers 160°.

2. Open grill and place a cheese slice on each burger.

3. Meanwhile, heat refried beans in microwave oven on high power (100%) for 30 to 45 seconds, or until heated through.

4. Place burger on bun. Divide refried beans between burgers. Top with salsa, guacamole, jalapeno peppers, and tortilla chips.

Makes 4 servings.
Preparation time: 5 to 10 minutes
Grill time: 5 to 8 minutes

Southwestern Bunkhouse Burgers

1 ½ pounds lean ground beef

1 package (1 ounce) chili seasoning

Salt, to taste

6 hamburger buns, split and toasted

6 slices cheddar cheese

6 slices tomato

Salsa

Tex-Mex Mayo (recipe following)

1. Preheat contact grill. Stir chili seasoning into beef, blending well; shape into 6 patties.

2. Grill patties for 5 to 8 minutes, or until meat thermometer registers 160°. Season cooked patties with salt, to taste.

3. Place cooked beef patty on a bun; top with cheese, tomato, salsa, and Tex-Mex Mayo.

Makes 6 servings.
Preparation time: 5 minutes
Grill time: 5 to 8 minutes

Tex-Mex Mayo

1 package (1 ounce) chili seasoning

1 ½ cups mayonnaise

⅓ cup chopped fresh cilantro

✦✦✦

1. Combine chili seasoning, mayonnaise, and cilantro in small bowl. Blend until combined.

2. Cover and refrigerate for several hours to blend flavors.

Tuna Burgers with Red Pepper Aioli

This is the perfect, healthy alternative to typical burgers. Adults will especially enjoy the complex flavor of these burgers, packed with vegetables and fresh tuna, then topped with a red pepper and mayonnaise sauce.

TIPS

Toasting the hamburger buns crisps the buns and also helps prevent the sauce from soaking into them.

¼ medium red pepper, cut into chunks

1 wedge red onion, about 1 inch thick

4 whole mushrooms

2 cloves garlic

1 slice fresh ginger, about ¼ inch thick

¾ pound fresh tuna steaks, or frozen tuna steaks, thawed and drained

1 tablespoon sesame oil

Salt and pepper, to taste

4 hamburger buns, toasted

Red Pepper Aioli (recipe following)

1. Place the red pepper, red onion wedge, mushrooms, garlic, and ginger in work bowl of food processor. Pulse to coarsely chop. Cut tuna into 1- to 2-inch pieces, then add to work bowl. Pulse to coarsely chop.

2. Add sesame oil, and season with salt and pepper; pulse to combine. (Do not overprocess; mixture should resemble ground beef in texture.) Shape into 4 patties.

3. Preheat contact grill. Grill patties for about 3 minutes, or until golden brown and cooked through. (Do not overcook.)

4. Serve on toasted hamburger buns, and top each with Red Pepper Aioli sauce.

Makes 4 servings.
Preparation time: 5 to 10 minutes
Grill time: 3 minutes

Red Pepper Aioli

TIPS

This sauce is good on any burger or sandwich, or it can be used as a sauce for grilled chicken or fish.

¼ cup roasted red peppers, drained

2 cloves garlic

½ cup mayonnaise

¼ teaspoon sugar

Dash salt

✳✳✳

1. Place red pepper and garlic in work bowl of food processor. Process until finely chopped.
2. Spoon into mixing bowl and add mayonnaise, sugar, and salt; stir to combine.

Turkey Burgers with Tomato-Ginger Relish

1 pound ground turkey

1 clove garlic, minced

1 teaspoon grated fresh ginger

½ teaspoon rubbed sage

½ teaspoon dried thyme leaves

¼ teaspoon salt

4 slices Muenster cheese

4 hamburger buns, toasted

Tomato-Ginger Relish (recipe following)

1. Place turkey in mixing bowl. Add garlic, ginger, sage, thyme, and salt. Mix well, and shape into 4 patties.

2. Preheat contact grill. Grill patties for 5 to 6 minutes, or until meat thermometer registers 165°.

3. Remove patties from grill, and top each immediately with a slice of cheese. Place on buns, and top each with Tomato Ginger Relish.

Makes 4 servings.
Preparation time: 5 minutes
Grill time: 5 to 6 minutes

Tomato-Ginger Relish

2 Roma tomatoes, seeded and chopped

1 green onion, chopped

1 tablespoon red wine vinegar

2 teaspoons soy sauce

1 teaspoon grated fresh ginger

1 clove garlic, minced

Dash crushed red pepper flakes

Salt, to taste

✦✦✦

1. Combine all ingredients. Allow to stand for 10 minutes before serving.

PIZZA

Chicken Bacon Club Pizza

This flavor of this unique pizza will remind you of a classic club sandwich.

TIPS

You can make your own pizza crust, following the recipe on page 263, or purchase a prepared pizza crust.

Substitute about 1 pound of thin turkey cutlet for chicken, if desired.

2 boneless, skinless chicken breast halves, pounded to about ½-inch thickness

4 strips bacon

¼ cup mayonnaise

½ teaspoon dried basil leaves

½ teaspoon dried thyme leaves

1 tablespoon mustard

1 prepared pizza crust, about 12 inches in diameter

1 medium tomato, seeded and chopped

1½ cups shredded provolone or mozzarella cheese

¼ cup shredded Parmesan cheese

½ cup finely shredded lettuce

✦ ✦ ✦

1. Preheat oven to 425°. Preheat contact grill. Cut bacon in half; arrange 2 half-pieces of bacon closely together on grill.

2. Place a chicken breast on top of bacon, covering bacon completely. Top chicken with 2 more half-pieces of bacon. Repeat with second chicken breast and remaining bacon.

3. Grill 5 to 6 minutes, or until chicken is fully cooked and bacon is crisp. Set aside until just cool enough to handle. Thinly slice chicken and bacon into strips.

4. Combine mayonnaise, basil, thyme, and mustard. Spread mayonnaise mixture over pizza crust. Top with cooked chicken and bacon, then tomato and cheeses.

5. Bake for 10 to 15 minutes, or until crust is golden and cheese is melted. Sprinkle with lettuce just before serving.

Makes 6 servings.
Preparation time: 15 minutes
Grill time: 5 to 6 minutes
Baking time: 10 to 15 minutes

Greek Mushroom Pizza

2 or 3 portobello mushrooms, sliced ½ inch thick

10 to 12 whole button mushrooms

2 tablespoons olive oil

2 cloves garlic, minced

½ cup prepared pizza sauce

1 teaspoon Greek seasoning blend or dried oregano leaves

1 prepared pizza crust, about 12 inches in diameter

4 thin slices prosciutto ham, cut into thin strips

⅓ cup finely chopped oil-packed sun-dried tomatoes

1 jar (6.5 ounces) marinated artichoke hearts, drained and chopped

¼ cup shredded Parmesan cheese

2 cups shredded pizza-blend cheese

1. Preheat oven to 450°. Place mushrooms in zip-top bag. Add olive oil and garlic; seal, toss gently to coat, and allow to stand for 5 minutes.

2. Preheat grill. Grill mushrooms for 3 to 5 minutes, or until tender. Cut button mushrooms in half. Mix pizza sauce with Greek seasoning; spread over crust.

3. Top with grilled mushrooms, prosciutto ham strips, finely chopped sun-dried tomatoes, and artichokes. Sprinkle cheese over all.

4. Bake for 10 to 12 minutes, or until cheese is melted and crust is brown.

Makes 6 servings.
Preparation time: 10 to 15 minutes
Grill time: 3 to 5 minutes
Baking time: 10 to 12 minutes

Grilled Chicken and Onion Pizza

TIPS

If you have caramelized onions in the freezer (see page 12), this would be a great time to use them, substituting them for the grilled onions in this recipe. Thaw ½ to ¾ cup of onions, arrange over pizza, and proceed as directed.

¾ pound boneless, skinless chicken breast tenders

1 tablespoon olive oil

1 clove garlic, minced

½ teaspoon Italian seasoning

1 large sweet onion, sliced ½ inch thick

1 tablespoon olive oil

1 prepared pizza crust, about 12 inches in diameter

1 tablespoon olive oil

⅓ cup ricotta

2 cloves garlic, minced

½ teaspoon dried basil leaves

½ teaspoon dried oregano leaves

1 cup sliced mushrooms

⅓ cup shredded Parmesan cheese

1 cup shredded mozzarella cheese

✦✦✦

1. Pound chicken to ½-inch thickness. Brush with 1 tablespoon of olive oil, then sprinkle evenly with 1 clove of minced garlic and Italian seasoning; set aside.

2. Preheat contact grill. Brush onion with olive oil. Grill onion slices for about 8 to 9 minutes, or until browned. Place onions in a bowl and cover with aluminum foil; allow to stand.

3. Grill chicken for about 5 to 6 minutes, or until meat thermometer registers 170°.

4. Preheat oven to 425°. Brush pizza crust with 1 tablespoon of olive oil. Combine ricotta, 2 cloves of garlic, basil leaves, and oregano leaves. Spoon ricotta mixture evenly over crust.

5. Slice chicken into strips about 1 inch thick. Top pizza with grilled chicken and mushrooms. Separate onion into rings, and arrange over chicken. Sprinkle with cheese.

6. Bake for about 15 minutes, or until cheese is melted.

Makes 4 to 6 servings.
Preparation time: 5 minutes
Grill time: 13 to 15 minutes
Baking time: 15 minutes

Grilled Vegetables and Caramelized Onion Pizza

TIPS

If desired, substitute grilled onions for caramelized onions. Slice 1 medium onion into ½-inch slices. Add to zip-top bag with other vegetables, and coat with garlic and olive oil. Grill onion slices for 7 to 10 minutes, or until tender.

½ red pepper, quartered

½ green pepper, quartered

½ small zucchini, sliced ½ inch thick

10 whole button mushrooms

3 cloves garlic, minced

3 tablespoons olive oil

1 prepared pizza crust, about 12 inches in diameter

¼ cup prepared basil pesto or prepared sun-dried tomato pesto

¼ cup pizza sauce

½ cup caramelized onions (see recipe, page 12)

2 Roma tomatoes, thinly sliced

¼ shredded Parmesan cheese

1½ cups shredded mozzarella or pizza-blend cheese

✦✦✦

1. Preheat oven to 450°. Slash bottom of peppers so they lay flat. Place peppers, zucchini, and mushrooms in zip-top bag. Add garlic and olive oil; seal, and shake to coat evenly. Allow to stand for 5 minutes.

2. Preheat grill. Grill peppers and zucchini for 7 to 10 minutes, or until tender. Grill mushrooms for 3 to 5 minutes, or until tender. Cut grilled mushrooms in half.

3. Spread pizza crust evenly with basil pesto and pizza sauce. Spoon caramelized onions over pizza crust. Arrange grilled vegetables and thinly sliced tomatoes over onions. Top with cheese.

4. Bake for 10 to 12 minutes, or until cheese is melted and crust is brown.

Makes 6 servings.
Preparation time: 10 to 15 minutes
Grill time: 10 to 15 minutes
Baking time: 10 to 12 minutes

Philly Steak Pizza

½ pound boneless beef sirloin steak, cut about ¾ inch thick

Salt and pepper, to taste

¼ teaspoon garlic powder

1 red or green pepper, quartered

1 medium onion, sliced ½ inch thick

2 tablespoons olive oil

1 prepared pizza crust, about 12 inches in diameter

½ cup pizza sauce

½ teaspoon minced garlic

1½ cups shredded cheddar cheese or Monterey Jack cheese

✛ ✛ ✛

1. Preheat oven to 450°. Preheat contact grill. Sprinkle steak with salt, pepper, and garlic powder.

2. Grill steak for 5 to 7 minutes, or until done as desired. Set steak aside and allow to stand for 5 minutes.

3. Slash peppers so they will lay flat. Place peppers and onions in zip-top bag. Add oil; seal, and toss to coat evenly. Grill peppers and onions for about 7 to 10 minutes, or until crisp-tender.

4. Spread pizza crust with sauce, then sprinkle with minced garlic. Thinly slice steak and peppers, and arrange over pizzas. Separate onion slices into rings and arrange over pizza. Sprinkle with cheese.

5. Bake for 10 to 12 minutes, or until cheese is melted and crust is brown.

Makes 6 servings.
Preparation time: 10 to 15 minutes
Grilling time: 12 to 17 minutes
Baking time: 10 to 12 minutes

Pizza Your Way

TIPS

You can divide the dough in half and make two 12-inch pizzas. Baking time will be the same.

It is easy to adapt the pizza topping recipes printed for a 12-inch pizza to smaller 8-inch pizzas. Most 8-inch pizza will require about ½ to ⅔ of the fillings suggested for a 12-inch pizza. Generally, don't overfill the pizza, for once it gets hot and the cheese is melted, the fillings will run.

Pizza crust is a great make-ahead food. Prepare the crust as directed, and bake just until set and firm, but not brown (about 5 minutes). Transfer to a rack and allow to cool completely. Wrap in plastic and freeze. Then, when ready to bake the pizza, transfer the

> *Yes, you can purchase the prebaked pizza crusts or focaccia that are so readily available now, but for a great taste treat, try your own pizza crust. It is easy to make, and the flavor is so great, you will not want to use the packaged versions again. Then, top that pizza any way you like it.*

1 package active dry yeast
1 teaspoon honey
1 cup warm water (105° to 115°F), divided
3¼ cups all-purpose flour
1 teaspoon salt
1 tablespoon olive oil

1. Dissolve yeast and honey in ¼ cup of warm water. Stir to blend well.

2. Use the dough hook of an electric mixer, and combine the flour and salt. Add olive oil, yeast mixture, and remaining ¾ cup of water. Mix on medium speed until the dough comes away from the sides of the bowl (about 5 to 7 minutes).

3. Remove the dough from the mixer bowl, and knead on a lightly floured surface for 3 to 4 minutes. Cover with a cloth and allow to rise in a warm place for 30 minutes.

4. Divide the dough into fourths. Form each fourth into a ball, pulling down the sides and tucking them under until you have a smooth, firm ball. Place on a baking sheet and cover with a damp towel; allow to rest for 15 to 25 minutes. (You may wrap each ball and refrigerate for up to 2 days.)

(continues on next page)

partially baked crust to a pizza pan, then top and bake until cheese is melted and edges of crust are golden.

5. Preheat a pizza stone or pizza pan in a 500° oven. On a lightly floured surface, stretch and form the ball into an 8-inch circle. Form a slightly raised edge along the outside.

6. Using a large spatula, gently and carefully transfer the dough to the pizza stone or pan. Carefully add toppings, and bake for 10 to 12 minutes.

Makes 4 pizzas, each about 8 inches in diameter.
Preparation time: 45 minutes
Baking time: 10 to 12 minutes

Pizza Toppings!

Use your contact grill and slow cooker to make great-tasting pizza toppings.

Slow Cooker

Prepare as directed, then spoon onto pizza crust.

Pizza Dip, see recipe on page 45

Caramelized Onions, see recipe on page 12

Marinara Sauce, see recipe on page 55

Grill

Meats, chicken, and seafood

Be sure meats are fully cooked before placing on pizza crust. Grill, then cut into thin strips or bite-size pieces.

Chicken

Italian or pork sausage links

Steaks

Shrimp

Vegetables

Slice vegetables about ½ inch thick; brush lightly with olive oil, and grill until done as desired.

Onions

Mushrooms

Zucchini or summer squash

Eggplant slices

Red, green, or yellow peppers

Asparagus

Pizza Creativity

Think of a pizza crust as an artist's palette, ready for you to top in any way you want. Experiment and try new flavors. Maybe this list will get you started thinking about new possibilities.

The amount of sauce, topping, and cheese will vary with the number of items you are using. Generally, about ¼ to ⅓ cup of sauce and ½ to 1 cup of topping or cheese will be plenty for an 8-inch pizza. For great family or party fun, make the pizza crusts above, then heat up the grill, grab caramelized onions you made previously in the slow cooker from the freezer, and set out a variety of these other toppings. Each person can then make his or her own signature pizza.

Sauces

Pizza or spaghetti sauce, pesto, Alfredo sauce, barbecue sauce, herb-flavored cream cheese, salsa

Hint: One jar (8 ounces) is generally enough topping for three (8-inch) pizzas

Herbs and Seasonings

Choose minced fresh herbs or dried seasonings! Try minced fresh basil, thyme, minced garlic, roasted garlic, diced minced jalapeno peppers, herb seasoning blends

Meats

Pepperoni, browned ground beef, browned pork sausage, Canadian bacon, smoked turkey, ham

Vegetables and Fruits

Fresh tomatoes, sliced pineapple, arugula, spinach, olives (try pitted kalamata, marinated Greek olives, and others from an olive bar), roasted red peppers, sun-dried tomatoes, artichoke hearts, and more

Cheese

Go beyond the shredded mozzarella and Parmesan, and add shredded cheddar, provolone, goat cheese, Asiago, Gorgonzola, fontina, and more

BEEF

Beef and Onion Fajitas

Fajitas are a delightful kind of family meal, since everyone can customize his or her own. Some will add grilled onions and pile it high with toppings. Others, especially the children, may choose only the beef to nestle in the tortilla. Either way, everyone is happy.

TIPS

Substitute chicken strips for beef, if desired. Grill until chicken is fully cooked and meat thermometer reaches 170° (about 5 to 6 minutes).

Be sure to slice the onion evenly. Uneven slices will hold the top of the grill up and cause uneven cooking.

1 pound beef flank steak

⅓ cup lime juice

⅓ cup vegetable oil

1 tablespoon Worcestershire sauce

1 tablespoon minced cilantro

1 jalapeno pepper, seeded and finely chopped

1 clove garlic, minced

½ teaspoon ground cumin

1 medium onion, sliced ¼ inch thick

Olive oil

Salt and pepper to taste

4 flour tortillas, about 8 inches in diameter, warmed

Pico de gallo or salsa, optional

Guacamole, optional

Sour cream, optional

+++

1. Slice flank steak, across the grain, into strips about 2 x ¼ inches in size. Place steak strips in a zip-top food bag.

2. Combine lime juice, vegetable oil, Worcestershire sauce, cilantro, jalapeno, garlic, and cumin; pour over steak. Seal bag and refrigerate several hours or overnight.

Makes 4 servings.
Preparation time: 5 to 10 minutes
Grill time: 7 to 8 minutes

3. Preheat contact grill. Drain steak and discard marinade. Brush onion slices with oil. Grill onion for 4 to 5 minutes, or until golden. Place onion in a bowl, cover with aluminum foil, and allow to stand.

4. Grill steak for about 3 minutes, or until done as desired. Place cooked steak in serving bowl and season to taste with salt and pepper.

5. To assemble, place cooked onions and steak on a flour tortilla; fold tortilla over meat and onions. Top, as desired, with pico de gallo, guacamole, and sour cream.

Caramelized Balsamic Onions on Sirloin

2 tablespoons butter

1½ pounds onions, thinly sliced

½ cup sugar

1 cup balsamic vinegar

1½ pounds boneless beef sirloin, cut about 1 inch thick

Salt and pepper, to taste

✦ ✦ ✦

1. Melt butter in large skillet over medium heat. Add the onions and cook slowly until very soft but not browned (about 12 minutes).

2. Blend in sugar and vinegar, and simmer until the liquid has almost all evaporated (about 35 minutes). Season with salt and pepper to taste. Serve warm, or at room temperature.

3. Meanwhile, preheat grill. Season steak with salt and pepper. Grill steak for about 6 to 8 minutes, or until desired doneness. Top grilled steak with caramelized balsamic onions.

Makes 6 servings.
Preparation time: 40 minutes
Grill time: 6 to 8 minutes

Flank Steak in Chimichurri Sauce

This dish is a favorite from Argentina. Give it a try!

¼ cup loosely packed fresh parsley

2 tablespoons loosely packed fresh oregano leaves

3 cloves garlic

¼ teaspoon salt

¼ teaspoon pepper

⅛ teaspoon crushed red pepper flakes

2 tablespoons fresh lemon juice

3 tablespoons white wine vinegar

¼ cup olive oil

1 pound beef flank steak

✦✦✦

1. Place parsley, oregano, garlic, salt, pepper, and red pepper flakes in work bowl of food processor. Pulse to chop. Add lemon juice, vinegar, and olive oil; process to combine.

2. Place steak in zip-top bag; pour half of herb-oil mixture over meat. Seal, and refrigerate for several hours or overnight. Cover and refrigerate remaining herb-oil mixture.

3. Preheat contact grill. Drain meat; discard marinade. Grill meat for 6 to 8 minutes, or until meat thermometer reaches 160° for medium doneness.

4. Slice meat into thin strips, cutting across the grain. Serve with reserved herb mixture.

Makes 6 servings.
Preparation time: 10 minutes
Grill time: 6 to 8 minutes

Garlic-Grilled Flank Steak

½ cup minced sweet onion

½ cup minced Italian parsley

3 cloves garlic, minced

¼ cup sherry vinegar

¼ cup olive oil

Dash hot pepper sauce

Salt and pepper, to taste

1 to 1½ pounds flank steak

✳ ✳ ✳

1. Combine onion, parsley, garlic, vinegar, olive oil, and pepper sauce; blend well, and set aside. Salt and pepper flank steak.

2. Preheat contact grill. Grill steak for 6 to 8 minutes, or until desired doneness. Place steak in glass dish; top with sauce, cover, and allow to stand for 5 to 10 minutes.

3. Slice steak very thinly, and return to glass dish. Spoon sauce over steak for serving.

Makes 4 servings.
Preparation time: 5 minutes
Grill time: 6 to 8 minutes

Grilled Marinated Roast

One might not think of grilling a roast. The marinade tenderizes the meat and gives us a quick, wonderful new way to serve that roast.

TIPS

This marinade does not have sugar. Since this thicker cut cooks a little longer than some, it is important to have a marinade that is not sweet, or it will brown too quickly. Grill the meat only to medium doneness, then cover it and allow it to stand for several minutes before slicing, so it will "finish cooking."

1 boneless beef arm roast, about 1 to 1 ½ pounds, cut about 1 ¼ to 1 ½ inches thick

⅓ cup vegetable oil

⅓ cup red wine vinegar

¼ cup soy sauce

1 teaspoon salt

1 teaspoon pepper

1 clove garlic, minced

✦✦✦

1. Place beef in zip-top bag. Combine remaining ingredients and pour over meat. Seal and refrigerate for several hours or overnight. Drain meat; discard marinade.

2. Preheat contact grill. Grill meat for about 15 minutes, or to medium doneness.

3. Remove and place on cutting board, cover with aluminum foil, and allow to stand for 10 minutes. Thinly slice meat, across the grain.

Makes 4 to 6 servings.
Preparation time: 5 minutes
Grill time: 15 minutes

Grilled Tenderloin with Pepper Cream Sauce

While you could save this dish just for company—why? It tastes so good, you will want to serve it often. Watch for a sale on those scrumptious beef tenderloins and treat your family.

1 tablespoon butter

1 clove garlic, minced

2 tablespoons cognac

1 teaspoon multicolored peppercorns, crushed

½ teaspoon dried oregano leaves

Pinch salt

⅔ cup heavy whipping cream

2 tablespoons sour cream

4 beef tenderloin fillets, about 6 ounces each

Salt and pepper, to taste

1. Melt butter in medium skillet over medium-high heat. Add garlic, and sauté for 1 minute.

2. Add cognac, peppercorns, oregano, and salt; simmer for 1 minute. Add whipping cream, and bring to a boil. Boil for 5 to 6 minutes, or until sauce is reduced by half. Blend in sour cream; keep sauce warm.

3. Preheat contact grill. Season steaks with salt and pepper. Grill for 5 to 6 minutes, or until desired doneness. Place on platter and drizzle with sauce.

Makes 4 servings.
Preparation time: 7 to 8 minutes
Grill time: 5 to 6 minutes

Indonesian-Style Grilled Steak

The foods of Indonesia are often thought of as spicy, exotic, and sometimes laced with peanuts. Give this unique dish a try!

TIPS

This recipe calls for one of the newer, lean cuts of the beef that is not normally thought of for grilling. Marinating overnight in this flavorful, rich marinade tenderizes the cut and adds flavor.

1 clove garlic, minced

2 tablespoons vegetable oil

1 tablespoon soy sauce

2 tablespoons lemon juice

$\frac{1}{4}$ cup creamy peanut butter

$\frac{1}{2}$ teaspoon sugar

$\frac{1}{2}$ teaspoon crushed red pepper

$\frac{1}{4}$ teaspoon ground cumin

$\frac{1}{4}$ teaspoon ground ginger

Salt and pepper, to taste

1 $\frac{1}{2}$ pounds boneless beef charcoal steak or tri-tip steak, cut about $\frac{3}{4}$ inch thick

✦✦✦

1. Combine all ingredients, except steak, in zip-top bag. Seal, and squeeze to blend thoroughly.

2. Cut steak into serving pieces, then place in peanut butter mixture. Seal, and squeeze to coat steak evenly. Refrigerate for several hours or overnight.

3. Preheat contact grill. Remove steak and discard marinade. Grill steak for 5 to 7 minutes, or until done as desired. Thinly slice, across the grain.

Makes 6 servings
Preparation time: 5 to
   10 minutes
Grill time: 5 to 7 minutes

Korean Barbecued Beef

The combination of sesame oil, honey, and crushed red pepper flakes gives a spicy, sweet, exotic flavor to this steak.

TIPS

This makes a beautifully rich, brown surface on the meat. It is best to cook to medium-rare, and then to cover and allow to stand; the meat will finish cooking as it stands.

1 to 1½ pounds beef flank steak

3 cloves garlic, minced

¼ cup soy sauce

2 tablespoons sesame oil

2 tablespoons honey

¼ teaspoon crushed red pepper flakes

✦ ✦ ✦

1. Place beef in zip-top bag. Combine garlic, soy sauce, sesame oil, honey, and red pepper flakes; pour over meat. Seal bag, and refrigerate for several hours or overnight.

2. Drain beef; discard marinade. Preheat contact grill. Grill for 6 to 8 minutes, or until medium-rare doneness.

3. Place on platter, cover with aluminum foil, and allow to stand for 5 to 10 minutes. Slice thinly, across the grain.

Makes 4 to 6 servings.
Preparation time: 5 minutes
Grill time: 6 to 8 minutes

Rib Eye Steak with Warm Tomato Salsa

This wonderful steak will taste like summer—all year round! In the summer, cut the fresh corn from the cob, but in the winter, go ahead and use frozen corn, and it will taste equally wonderful.

TIPS

The spicy heat of a jalapeno pepper is found in its seeds and membranes, so a seeded jalapeno pepper will be more mildly flavored.

2 rib eye steaks, about 10 ounces each, cut about 1 inch thick

Coarse salt and coarsely ground pepper, to taste

2 tablespoons butter

1 medium onion, chopped

2 cloves garlic, minced

$\frac{1}{2}$ teaspoon chili powder

2 cups fresh or frozen corn

1 jalapeno pepper, seeded and finely chopped

$1\frac{1}{2}$ cups grape tomatoes, halved

$\frac{1}{4}$ cup chopped fresh basil

1 tablespoon freshly squeezed lime juice

✦✦✦

1. Preheat contact grill. Grill steaks for 6 to 8 minutes, or until cooked as desired. When steaks are done, remove to a platter and cover with aluminum foil; allow to stand for 5 to 10 minutes.

2. Meanwhile, prepare salsa. Melt butter in heavy, medium-size skillet. Add onion and cook, stirring frequently, over medium-high heat until onion is golden.

3. Add garlic and chili powder, and cook for 1 minute. Stir in corn and pepper. Cover and cook until corn is tender—about 3 minutes if fresh corn, or 5 minutes if frozen.

4. Add tomatoes and cook just until tomatoes soften. Remove from heat, and stir in basil and lime juice. Slice steaks across the grain, and serve with warm tomato salsa.

Makes 4 servings.
Preparation time: 10 minutes
Grill time: 6 to 8 minutes

Sizzling Steak

TIPS

This marinade would be great on any cut of beef steak.

> *This steak is packed with flavor, but it is not as hot as you might think. Go ahead and try it. The aroma alone is almost hypnotic.*

1 package (1 ¼ ounces) taco seasoning

1 can (12 ounces) beer, at room temperature

2 tablespoons hot pepper sauce

4 rib eye steaks, cut about 1 inch thick

Salt and pepper, to taste

✦✦✦

1. Combine taco seasoning, beer, and hot pepper sauce; stir well to blend. Place steak in a zip-top bag and pour beer mixture over to coat. Seal bag and refrigerate for several hours or overnight.

2. Preheat contact grill. Drain meat; discard marinade. Grill steaks for about 6 to 8 minutes for medium doneness, or until done as desired. Season with salt and pepper.

Makes 4 servings.
Preparation time: 5 minutes
Grill time: 6 to 8 minutes

Southwestern Steak and Potato Salad

1 package (1 ¼ ounces) taco seasoning

¼ cup lime juice

¼ cup olive oil

1 ½ pounds boneless top sirloin steak

1 ½ pounds small red potatoes, quartered

⅓ cup mayonnaise

6 cups torn romaine lettuce

2 cups halved cherry tomatoes

½ cup chopped celery

3 tablespoons chopped cilantro

½ cup sliced ripe olives

✦ ✦ ✦

1. In a small bowl, whisk together the taco seasoning, lime juice, and olive oil. Pour about ¼ cup of mixture into a zip-top bag. Add steak, seal bag, and refrigerate for several hours. Cover remaining seasoning mixture and refrigerate for use as salad dressing.

2. Preheat contact grill. Drain meat, discarding marinade. Grill meat for 5 to 8 minutes, or until done as desired. Thinly slice steak.

3. Meanwhile, cook potatoes in boiling water until tender, about 8 to 10 minutes; drain. For salad dressing, whisk together reserved seasoning mixture and mayonnaise in a small bowl.

4. Combine romaine, cooked potatoes, and remaining ingredients in salad bowl. Drizzle with dressing and toss to combine. Top with steak strips.

Makes 4 servings.
Preparation time: 15 minutes
Grill time: 5 to 8 minutes

Steak, Bistecca Style

TIPS

Substitute other fresh herbs, such as minced fresh rosemary or thyme, for the basil.

> *We usually don't think of steak—bistecca—when we think of the great foods of Italy. Yet, they have perfected cooking succulent steaks. This strip steak is cooked and seasoned in a similar fashion to those you might enjoy in the Tuscan countryside.*

2 boneless beef strip steaks, about 10 to 12 ounces each, and cut 1¼ inches thick

1 clove garlic, cut in half

2 tablespoons extra virgin olive oil

Salt

Freshly ground black pepper

2 tablespoons balsamic vinegar

2 tablespoons minced Italian-style parsley

1 tablespoon minced fresh basil or oregano

✦ ✦ ✦

1. Rub garlic on both sides of the steaks. Brush both sides of the steaks with olive oil. Sprinkle generously with salt and pepper.

2. Preheat contact grill. Grill steaks for 6 to 8 minutes, or until meat thermometer registers 145° for medium-rare, or until done as desired. Remove steaks, cover, and allow to stand for 5 minutes.

3. Thinly slice steaks across the grain, and arrange on serving plate. Drizzle with balsamic vinegar, then sprinkle with fresh herbs.

Makes 4 servings.
Preparation time: 3 to 5 minutes
Grill time: 6 to 8 minutes

PORK AND LAMB

Balsamic Glazed Pork

This dish is so simple—but oh, so tasty.

¼ cup olive oil

⅓ cup balsamic vinegar

4 center-cut boneless pork chops, about 1 pound

Salt and pepper, to taste

✶✶✶

1. Combine oil and vinegar in small bowl. Place pork in zip-top bag; pour marinade over all and allow to stand for 30 minutes.

2. Preheat contact grill. Drain chops and discard marinade. Grill chops for about 5 minutes, or until meat thermometer registers 160° for medium doneness, or until done as desired. Season to taste with salt and pepper.

Makes 4 servings.
Preparation time: 5 minutes
Grill time: 5 minutes

Grilled Ham Cobb Salad

Makes 3 to 4 servings.
Preparation time: 15 minutes
Grill time: 2 to 3 minutes

½ to ¾ pound boneless ham steak, cut about ¾ to 1 inch thick

1 tablespoon Dijon mustard

6 to 8 cups torn lettuce, or 1 package (10 to 12 ounces) prepared salad mix

2 green onions, sliced

1 large tomato, seeded and chopped

⅓ cup ripe, pitted olives, sliced

2 hard-cooked eggs, sliced

3 slices bacon, cooked until crisp and crumbled

½ cup shredded Swiss or cheddar cheese

½ cup ranch dressing

1. Preheat contact grill. Brush ham steak evenly with mustard. Grill for 2 to 3 minutes, or until steak is hot and meat thermometer registers 140°. Set aside.

2. Combine lettuce, green onions, tomato, and olives in salad bowl; toss to combine. Slice ham into thin strips, and arrange over lettuce. Top with sliced egg, crumbled bacon, and cheese. Drizzle with dressing.

Grilled Lamb with Pine-Nut Rice Pilaf

Lamb is a great change of pace from other dinners. This lamb is flavored with mint, as well as cumin, oregano, and lemon, for a delightful taste.

TIPS

Toasting the pine nuts intensifies their flavor. To toast pine nuts, spread them in a shallow baking pan. Bake at 350° for about 6 to 8 minutes, or until golden.

The pine-nut rice pilaf is great with other meat dishes, too. Serve with pork roast or grilled chicken, if desired.

1 tablespoon butter

1 clove garlic, minced

¾ cup uncooked long-grain rice

1½ cups water

¼ teaspoon salt

2 tablespoons vegetable oil

Grated zest of 1 lemon

2 teaspoons dried oregano leaves

1 tablespoon minced fresh mint

¼ teaspoon ground cumin

½ teaspoon salt

1 pound boneless lamb, cut about 1 inch thick

6 to 8 whole mushroom caps

1 small onion, sliced

2 tablespoon vegetable oil

⅓ cup pine nuts, toasted

2 tablespoons minced fresh mint

✦ ✦ ✦

1. Melt butter in a saucepan over medium-high heat. Add garlic, and sauté for about 30 seconds.

2. Stir in rice and cook, stirring frequently, until rice is golden brown (about 2 to 3 minutes). Stir in water and ¼ teaspoon of salt. Cover, reduce heat, and simmer for 20 minutes, or until water is absorbed.

Makes 4 servings.
Preparation time: 20 to 25 minutes
Grill time: 11 minutes

3. Combine 2 tablespoons of oil, lemon zest, oregano, 1 tablespoon of mint, cumin, and $\frac{1}{2}$ teaspoon of salt; rub over both sides of lamb.

4. Preheat contact grill. Brush mushrooms and onion slices with 2 tablespoons of oil. Grill vegetables for about 5 minutes, or until browned. Remove and place in a bowl; cover with aluminum foil, and allow to stand for 5 minutes.

5. Grill lamb for 6 minutes, or until meat thermometer registers 160° for medium doneness.

6. Chop grilled mushrooms and onions. Stir into cooked rice. Stir in pine nuts and remaining 2 tablespoons of fresh mint. Serve lamb with rice.

Jamaican Jerk Pork

1 pound pork tenderloin, cut into slices about 1 inch thick

1 cup pineapple juice

¼ cup olive oil

2 cloves garlic, minced

2 teaspoons jerk seasoning

¼ teaspoon salt

1 can (8 ounces) crushed pineapple in juice

⅓ cup white wine

¼ cup brown sugar

⅓ cup finely chopped red onion

1 teaspoon minced fresh ginger

1 jalapeno pepper, seeded and minced

6 to 8 fresh spinach leaves

½ cup thinly sliced red pepper strips, optional

½ cup thinly sliced, peeled jicama strips, optional

✦ ✦ ✦

1. Place pork tenderloin in zip-top bag. Add pineapple juice, olive oil, 2 cloves of garlic, jerk seasoning, and salt; seal bag, and refrigerate overnight.

2. Place crushed pineapple, wine, brown sugar, red onions, ginger, and jalapeno pepper in small saucepan. Heat to a boil; reduce heat and simmer for 10 to 15 minutes, or until thickened, stirring occasionally. Cool, cover, and refrigerate overnight.

3. Preheat contact grill. Drain pork, and discard marinade. Grill for 7 to 8 minutes, or until meat thermometer registers 160°. Season pork with salt and pepper.

4. To serve, arrange spinach leaves on serving plate. Top with grilled pork and pineapple salsa. Garnish each serving with red pepper and jicama strips, if desired.

Makes 4 servings.
Preparation time: 20 to 25 minutes
Grill time: 7 to 8 minutes

Margarita Chops

This marinade adds a great new flavor to pork chops. Another time, try the marinade on boneless chicken breasts.

4 boneless pork chops, cut about ¾ to 1 inch thick

2 tablespoons lime juice

2 tablespoons tequila

2 tablespoons vegetable oil

½ teaspoon cumin

½ teaspoon salt

½ teaspoon pepper

✳ ✳ ✳

1. Place pork chops in zip-top bag. Combine remaining ingredients, and pour over meat. Seal bag, and refrigerate for several hours or overnight.
2. Preheat contact grill. Drain chops, and discard marinade. Grill chops for about 5 minutes, or until meat thermometer registers 160°.

Makes 4 servings.
Preparation time: 5 minutes
Grill time: 5 minutes

Mustard-Glazed Ham Steak

TIPS

The round bone in a ham steak is easily removed before cooking. Bones, if left in the meat, will not allow the grill to close and will cause uneven cooking.

2 tablespoons spicy brown mustard

2 tablespoons lemon juice

1 tablespoon Worcestershire sauce

1 tablespoon brown sugar

$\frac{1}{4}$ teaspoon ground cloves

1 ham steak, about 1$\frac{1}{2}$ to 1$\frac{3}{4}$ pounds, cut about 1 inch thick

✴✴✴

1. Preheat contact grill. Combine mustard, lemon juice, Worcestershire sauce, brown sugar, and cloves.

2. If ham steak has a bone in it, cut around bone and remove it. Brush both sides of ham steak with mustard glaze.

3. Grill 3 to 5 minutes, or until meat thermometer registers 140°. Serve steak with remaining glaze.

Makes 4 to 6 servings.
Preparation time: 3 to
5 minutes
Grill time: 3 to 5 minutes

Pesto Pocket Chops

TIPS

Placing the chops in the freezer until they are icy, but not solidly frozen, makes them easier to slice.

These chops are especially good served with Basil Cream Sauce (see recipe, page 296).

4 thick, boneless pork chops, cut about 1 ¼ to 1 ½ inches thick

¼ cup pesto (see recipe, page 10)

2 tablespoons olive oil

Salt and pepper, to taste

✦ ✦ ✦

1. Place pork chops in freezer until just icy. With a sharp knife, carefully slice a pocket inside each chop, cutting horizontally from one of the outer edges. Spread pesto inside pocket.

2. Brush outside of each chop with olive oil, then season with salt and pepper. Allow to stand for 5 minutes.

3. Preheat contact grill. Grill chops for 7 to 8 minutes, or until meat thermometer registers 160°.

Makes 4 servings.
Preparation time: 5 minutes
Grill time: 7 to 8 minutes

Pork Souvlaki

Yes, the name sounds mysterious, but the flavors aren't. The name refers to a classic Greek dish, often prepared with lamb, and flavored with garlic and oregano.

1 pound boneless pork loin, cut into ¾-inch cubes

¼ cup olive oil

3 cloves garlic, minced

1 teaspoon dried oregano leaves

Salt and pepper, to taste

4 pita or flat breads

4 slices onion, each about ¼ inch thick

Olive oil

½ cup Yogurt Sauce (see page 243)

¾ cup chopped cucumber

2 tablespoons minced fresh parsley

1. Place pork cubes in zip-top bag. Combine ¼ cup of olive oil, garlic, oregano, salt, and pepper; pour over pork. Seal and refrigerate for several hours or overnight.

2. Lightly brush pita and onion slices with olive oil. Preheat contact grill. Grill pitas, one at a time, for about 1 minute or until hot. Wrap warm pita in aluminum foil to keep warm.

3. Grill onion slices until golden and tender, for about 5 minutes. Remove onions from grill; cover to keep warm.

4. Drain pork; discard marinade. Thread pork cubes onto 6- to 8-inch skewers. Grill pork skewers for about 6 minutes, or until pork is fully cooked.

5. To serve, top warm pita with a grilled onion slice. Top each with cooked pork. Stir together Yogurt Sauce and chopped cucumbers, and dollop over pork. Sprinkle with minced parsley.

Makes 4 servings.
Preparation time: 5 to 10 minutes
Grill time: 12 minutes

Quick Ginger Pork

1 pork tenderloin, about 1½ pounds

2 tablespoons soy sauce

2 tablespoons rice wine vinegar

2 tablespoons vegetable oil

1 teaspoon ground ginger

1 teaspoon garlic powder

Salt and pepper, to taste

✳✳✳

1. Slice pork into thick slices, about 1½ to 2 inches thick. Combine remaining ingredients and pour into a zip-top bag. Add pork, seal, and turn to coat evenly. Allow to stand at room temperature for about 10 minutes.

2. Preheat contact grill. Drain meat and discard marinade. Grill for about 6 minutes, or until meat thermometer registers 160°.

Makes 4 servings.
Preparation time: 5 minutes
Grill time: 6 minutes

Rosemary-Peppered Tenderloin

Our favorite way to prepare this dish is outside, on the barbecue grill. Yet, in the dead of winter, when snow, ice, or cold winds blow, the contact grill cooks a tasty meal.

TIPS

To make on the BBQ grill: Allow coals to burn down to white ash. Soak toothpicks in water for 10 minutes, to prevent burning. Do not cut tenderloin in half; otherwise, prepare as directed, wrapping both slices of bacon around tenderloin. Grill tenderloin over medium-hot coals, for about 25 to 30 minutes, turning every 10 minutes to brown evenly.

1 pound pork tenderloin

2 tablespoons Dijon mustard

1 teaspoon lemon pepper

1 teaspoon dried rosemary leaves, crushed

2 slices bacon

✣✣✣

1. Cut tenderloin in half to make 2 equal-sized pieces, and set aside. Combine mustard, pepper, and rosemary; spread mustard mixture evenly over both tenderloin pieces.

2. Spiral bacon slices around tenderloin, and secure ends of bacon with toothpick. (Be sure to push toothpick deeply into meat so only tips of toothpick show, to avoid burning toothpick.)

3. Preheat contact grill. Grill for 7 to 9 minutes, or until meat thermometer registers 160°.

4. Remove meat from grill and place on platter; cover with aluminum foil, and allow to rest for 5 to 10 minutes. Remove toothpicks, slice and serve.

Makes 4 servings.
Preparation time: 5 to 10 minutes
Grill time: 7 to 9 minutes

Spicy Asian Chops on Lo Mein

TIPS

Asian chili sauce with garlic purée is a spicy condiment found in the Asian food section of larger grocery stores.

This is one time when it is great to visit the salad bar of the grocery store and select 2 cups of a variety of prepared vegetables.

Makes 4 servings.
Preparation time: 10 to 15 minutes
Grill time: 5 to 6 minutes

¼ cup apricot preserves

1 tablespoon Asian chili sauce with garlic purée

1 tablespoon soy sauce

1 teaspoon sesame oil

4 boneless pork chops, cut about 1 inch thick

Salt and pepper, to taste

1 tablespoon vegetable oil

2 cups chopped vegetables, such as celery, mushrooms, green onions, carrots

2 tablespoons soy sauce

1 tablespoon sesame oil

Pinch sugar

8 ounces fettuccine noodles, cooked according to package directions, drained

1 tablespoon sesame seeds, toasted

✦✦✦

1. Combine preserves, chili sauce, 1 tablespoon of soy sauce, and 1 teaspoon of sesame oil. Measure out and reserve 2 tablespoons of mixture.

2. Preheat contact grill. Season chops with salt and pepper. Grill chops for 2 minutes.

3. Brush with preserves mixture. Continue grilling for 3 to 4 minutes, or until meat thermometer registers 160°.

4. Meanwhile, heat vegetable oil in large skillet over medium-high heat. Add vegetables and cook, stirring frequently, until vegetables are crisp-tender. Remove from heat. Stir in 2 tablespoons of soy sauce, 1 tablespoon of sesame oil, and sugar. Add cooked fettuccine, and toss to coat.

5. Arrange Lo Mein in deep serving platter. Top with cooked chops. Drizzle with reserved 2 tablespoons of apricot sauce, and sprinkle with sesame seeds.

SAUCES AND MARINADES

Basil Cream Sauce

This rich, flavorful sauce is great on grilled chicken or pork chops. If desired, serve with a side of hot cooked pasta.

_____
TIPS
_____

If desired, substitute ⅓ cup of finely chopped onion for shallot.

¼ cup white wine

1 shallot, chopped

2 cloves garlic, minced

½ cup heavy whipping cream

1 can (14.5 ounces) diced tomatoes

¼ cup fresh minced basil

Salt and pepper, to taste

✦✦✦

1. Combine wine, shallot, and garlic in small saucepan. Heat to a boil, then reduce heat. Cook, uncovered, for 2 to 3 minutes, or until vegetables are tender and wine has nearly evaporated.

2. Stir in cream, tomatoes, and basil. Simmer, stirring occasionally, for 5 minutes. Season to taste with salt and pepper.

Makes approximately 2 cups.
Preparation time: 3 minutes
Cooking time: 7 to 8 minutes

Chili-Cumin Butter

This versatile seasoned butter is great on grilled beef, pork, chicken, or fish.

½ cup butter, softened
1 teaspoon chili powder
½ teaspoon dried oregano leaves
½ teaspoon Worcestershire sauce
¼ teaspoon cumin
¼ teaspoon cayenne pepper
Dash garlic powder
Dash salt

1. Mix together all ingredients. Shape into a log, then wrap in plastic and refrigerate.
2. When ready to serve, slice into thin pieces and place over hot, grilled meat or fish.

Makes ½ cup.
Preparation time: 2 to
3 minutes

Cilantro Butter

> *Transform a plain grilled chicken breast, pork chop, or fish fillet into a flavorful dish with a spoonful of this flavored butter.*

½ cup butter, softened

⅓ cup fresh cilantro, minced

1 clove garlic, minced

2 tablespoons lime juice

Salt and pepper, to taste

✦ ✦ ✦

1. Mix together all ingredients. Shape into a log, then wrap in plastic and refrigerate.
2. When ready to serve, slice into thin pieces and place over hot, grilled meat or fish.

Makes ½ cup.
Preparation time: 2 to
   3 minutes

Dill Butter

Dab a bit of this seasoned butter over freshly grilled fish or chicken for a flavorful finish!

TIPS

A seasoned butter can be made ahead of time; just wrap tightly and refrigerate for up to one week.

For the best flavor, begin with real butter instead of margarine.

½ cup butter, softened

2 tablespoons minced fresh dill

1 tablespoon lemon juice

Salt and pepper, to taste

✦ ✦ ✦

1. Mix together all ingredients. Shape into a log, then wrap in plastic and refrigerate.
2. When ready to serve, slice into thin pieces and place over hot, grilled fish or chicken.

Makes ½ cup.
Preparation time: 2 to
   3 minutes

Mushroom Wine Sauce

This sauce is a classic and makes any meat special. Serve it over grilled beef, pork, or chicken. It is also great over roast beef cooked in the slow cooker.

3 tablespoons butter

1 cup sliced fresh mushrooms

¼ cup finely chopped onion

3 tablespoons all-purpose flour

1 cup milk

2 tablespoons soy sauce

¼ cup white wine

Salt and freshly ground black pepper

1. Melt butter in saucepan over medium heat. Add mushrooms and onion; cook, stirring frequently, until vegetables are tender.

2. Add flour, stirring to form a smooth paste. Cook, stirring, for 1 minute. Stir in milk and soy sauce. Cook, stirring constantly, until mixture bubbles and thickens.

3. Stir in wine, and allow to heat through. Season to taste with salt and pepper.

Makes 1¾ cup.
Preparation time: 5 to
    10 minutes
Cooking time: 5 minutes

Orange Balsamic Sauce

This sauce is great over grilled chicken, pork, or fish.

1 tablespoon cornstarch

1 cup chicken broth

2 tablespoons balsamic vinegar

2 teaspoons brown sugar

2 teaspoons orange juice

Salt and pepper, to taste

1. Blend together cornstarch and chicken broth in a small saucepan.
2. Stir in remaining ingredients. Heat to a boil, stirring constantly. Cook, stirring, until slightly thickened and bubbly.

Makes about 1 cup.
Preparation time: 5 minutes

Pineapple Soy Marinade

TIPS

If you enjoy this marinade, keep the small, individual-size cans of pineapple juice on hand in the pantry.

Use this marinade on boneless chicken breasts.

½ cup soy sauce

½ cup pineapple juice

⅓ cup red wine vinegar

⅓ cup vegetable oil

2 cloves garlic, minced

Salt and pepper, to taste

✳✳✳

1. Combine all ingredients. Pour over chicken, refrigerate, and marinate about 30 minutes.

2. Drain and grill chicken as directed.

Makes 1⅔ cups.
Preparation time: 3 minutes

Snappy Lemon Marinade

Use this marinade on beef, chicken, or fish. Best of all, this crisp marinade also makes a wonderful dressing for fresh green salad.

½ cup freshly squeezed lemon juice

2 tablespoons sherry vinegar

¾ cup olive oil

1 teaspoon minced garlic

½ teaspoon kosher salt

1 ½ teaspoons freshly ground black pepper

✦ ✦ ✦

1. Whisk all ingredients together.

Makes about 1⅓ cups.
Preparation time: 3 minutes

Sweet Mahogany Sauce

This is a fantastic sweet, yet spicy, dipping sauce. Serve with any grilled beef or pork, or brush it over a grilled burger.

¼ cup ketchup

2 tablespoons molasses

1 teaspoon Worcestershire sauce

1 teaspoon Dijon mustard

2 cloves garlic, minced

½ teaspoon hot pepper sauce

¼ teaspoon salt

¼ teaspoon pepper

✳ ✳ ✳

1. Combine all ingredients in a microwave-safe bowl.
2. Cover and microwave on high power (100%) for 30 seconds, or until hot.

Makes about ½ cup.
Preparation time: 5 minutes

CHICKEN AND POULTRY

Bacon-Wrapped Stuffed Grilled Chicken

This is a huge hit at our dinner table. I have to double the recipe for our family of three! My husband can't get over how moist the chicken is, and our daughter has seconds and then thirds.

TIPS

Be sure to keep chicken breasts whole for this recipe. This may be one time when you will need to purchase the chicken from the service meat counter and avoid the packaged selections.

2 whole boneless, skinless chicken breasts

4 to 6 tablespoons light garlic-and-herb cheese spread

4 slices bacon

✦✦✦

1. Leave each breast whole, and pound to $\frac{1}{4}$-inch thickness. Spread 2 to 3 tablespoons cheese spread over one side of each breast. Fold breast over. Using toothpicks, secure bacon around chicken to wrap completely.

2. Preheat contact grill. Grill chicken for 8 minutes, or until chicken is fully cooked and meat thermometer registers 170°. Cut each piece in half and serve.

Makes 4 servings.
Preparation time: 5 minutes
Grill time: 8 minutes

Chicken Spiedini

4 boneless, skinless chicken breast halves, about 1 pound

$\frac{2}{3}$ cup Italian-seasoned bread crumbs

$\frac{1}{3}$ cup grated Parmesan cheese

1 tablespoon chopped fresh Italian parsley

2 teaspoons grated lemon zest

1 tablespoon minced garlic

$\frac{1}{4}$ teaspoon crushed red pepper flakes

2 tablespoons butter or margarine, melted

2 tablespoons extra virgin olive oil

1 tablespoon freshly squeezed lemon juice

✦ ✦ ✦

1. Pound chicken until about $\frac{1}{4}$ inch thick. In shallow plate, combine bread crumbs, cheese, parsley, lemon zest, garlic, and red pepper flakes. In another shallow dish, combine butter and olive oil.

2. Dip each chicken piece in butter mixture; then place in crumb mixture and coat both sides evenly. Tightly roll up each piece, and cut into 1-inch slices. Place on skewers.

3. Preheat contact grill. Grill for 5 minutes, or until meat thermometer registers 170°. Place on serving platter and sprinkle lemon juice evenly over cooked chicken.

Makes 4 servings.
Preparation time: 10 minutes
Grill time: 5 minutes

Fiesta Cobb Salad

½ cup mayonnaise

½ cup sour cream

⅓ cup bottled Italian vinaigrette salad dressing

1 package (1¼ ounces) taco seasoning mix

4 boneless, skinless chicken breast halves

1 package (10 ounces) salad greens

1½ cups frozen corn, thawed and drained

1 avocado, peeled and diced

1 can (15 ounces) black beans, rinsed and drained

1 can (14.5 ounces) diced tomatoes and green chiles, drained

1 cup shredded Mexican-blend cheese

Tortilla chips

✛ ✛ ✛

1. Preheat contact grill. In a small mixing bowl, combine mayonnaise, sour cream, Italian dressing, and taco seasoning.

2. Pound chicken to ½-inch thickness and place in zip-top bag. Pour ¼ cup of dressing mixture over chicken; seal bag, and turn to coat. Cover and set remaining dressing mixture aside.

3. Grill chicken for 5 to 6 minutes, or until meat thermometer registers 170°. Allow chicken to cool for 5 minutes; cut into bite-size cubes.

4. Toss together salad greens with ½ cup of dressing. Divide greens onto 4 serving plates. Arrange the corn, avocado, chicken, beans, tomato, and cheese in rows over greens. Garnish with tortilla chips, and serve with remaining dressing.

Makes 4 servings.
Preparation time: 10 minutes
Grill time: 5 to 6 minutes

Fusilli with Grilled Chicken, Lemon, and Mushrooms

This tastes great reheated the next day for lunch. This pasta dish is light and full of flavor.

1 pound boneless, skinless chicken breast halves

Olive oil

Salt and freshly ground black pepper, to taste

12 ounces fusilli

3 tablespoons olive oil

1 pound button mushrooms, sliced

2 cloves garlic, minced

1 tablespoon grated lemon peel

⅓ cup freshly squeezed lemon juice

½ cup chopped fresh basil

½ teaspoon crushed red pepper flakes

Salt and pepper to taste

⅔ cup freshly grated Parmesan cheese

1. Pound chicken until thin. Brush with olive oil; season with salt and pepper.

2. Preheat contact grill. Grill chicken for 5 to 6 minutes, or until meat thermometer reaches 170°. Remove chicken from grill. Place on plate, then cover with foil and allow to rest. Set aside.

3. Cook fusilli according to package directions. Drain well and return to pot.

4. Meanwhile, heat 3 tablespoons of olive oil in large skillet. Add mushrooms and garlic; cook over medium-high heat for 3 minutes.

5. Slice chicken into strips and add to mushrooms. Add chicken mixture, and remaining ingredients, to fusilli. Toss together lightly. Garnish with fresh, grated Parmesan cheese.

Makes 6 servings.
Preparation time: 15 minutes
Grill time: 5 to 6 minutes

Grilled Chicken and Vegetables with Pesto Pasta

This is a family favorite, and one that the kids are sure to enjoy, too. Pasta is everyone's favorite, and the pesto flavor helps "disguise" those healthy veggies that are part of this great dish.

TIPS

Save about ½ cup of water from cooking pasta; use the reserved water to thin pesto, if needed.

This is a one-dish platter meal. All you need to add is a salad and crusty bread; then, enjoy a glass of wine.

4 boneless, skinless chicken breast halves

3 tablespoons freshly squeezed lemon juice

¼ cup extra virgin olive oil

2 cloves garlic, minced

2 yellow squash or zucchini, sliced about ½ inch thick

Olive oil

1 cup grape tomatoes, cut in half

½ cup fresh basil, cut into thin strips

12 ounces penne, cooked according to package directions

¾ cup basil pesto (see recipe, page 10)

✦ ✦ ✦

1. Pound chicken to ¼-inch thickness and place in zip-top bag. Combine lemon juice, ¼ cup of olive oil, and garlic; pour over chicken. Seal, and allow to marinate for 15 minutes.

2. Brush squash with olive oil. Preheat contact grill. Grill squash until tender. Allow to cool slightly, then cut into cubes. Place cooked squash in a bowl, and add tomatoes and basil. Cover and set aside.

3. Drain chicken; discard marinade. Grill chicken for about 5 to 6 minutes, or until meat thermometer registers 170°.

4. Toss cooked pasta with pesto. Spoon into large serving platter. Spoon vegetables over pasta, and top with cooked chicken.

Makes 4 servings.
Preparation time: 15 minutes
Grill time: 10 to 11 minutes

Grilled Chicken Chinese Salad

No need to go out to a restaurant! Get that restaurant flavor quickly at home.

Marinade:

3 tablespoons soy sauce

1 tablespoon rice vinegar

1 tablespoon sesame oil

1 clove garlic, minced

2 teaspoons minced fresh ginger

2 chicken breast halves

Dressing:

3 tablespoons sesame oil

1 tablespoon canola oil

3 tablespoons rice vinegar

1 tablespoon sugar

1 tablespoon soy sauce

1 bag (10 ounces) romaine, cabbage, and carrot salad blend, or 6 cups torn lettuce

3 green onions, sliced thin

¼ cup dry roasted cashews

½ cup chow mein noodles

✳✳✳

1. For marinade: Combine 3 tablespoons of soy, 1 tablespoon of rice vinegar, 1 tablespoon of sesame oil, garlic, and ginger in a shallow dish.

2. Pound chicken to about ½-inch thickness. Add chicken, and turn to coat. Allow to marinate for 15 minutes.

Makes 4 to 6 servings.
Preparation time: 10 minutes
Grill time: 5 to 6 minutes

(continues on next page)

3. Preheat contact grill. Remove chicken from marinade; discard marinade. Grill chicken for 5 to 6 minutes, or until meat thermometer registers 170°.

4. For dressing: Combine 3 tablespoons of sesame oil, canola oil, 3 tablespoons of rice vinegar, sugar, and 1 tablespoon of soy. Blend well and set aside.

5. Place romaine or lettuce in large salad bowl. Toss with green onions. Cut chicken into pieces and add to salad along with cashews. Add dressing to salad, and toss. Top with chow mein noodles.

Grilled Chicken Parmesan

This is really quick to prepare and is truly one of those dishes that tastes great. Better yet, you can have a home-cooked meal more quickly than loading the family into the car and buying dinner in a bag from a drive-through.

TIPS

Substitute 3 cups of Marinara Sauce (page 55) for the jar of Marinara Sauce.

To streamline the work and have everything done and ready to serve at the same time, first put on the water to boil for the pasta, then turn on the oven and proceed with the recipe.

This is one time when you should select smaller chicken breasts, each weighing 4 to 6 ounces. With the sauce and cheese, the servings will still be wonderfully filling, but the smaller chicken breasts will all fit on the contact grill at one time. If you choose larger ones, you may have to cook them in batches, and the dish will take a little longer to prepare.

Makes 4 servings.
Preparation time: 5 minutes
Grill time: 4 to 5 minutes
Baking time: 15 minutes

4 boneless, skinless chicken breast halves

1 tablespoon olive oil

$\frac{1}{2}$ teaspoon garlic powder

$\frac{1}{2}$ teaspoon dried basil leaves

$\frac{1}{4}$ teaspoon salt

$\frac{1}{4}$ teaspoon pepper

$\frac{1}{4}$ cup grated Parmesan cheese

1 jar (28 ounces) Marinara Sauce

$\frac{1}{2}$ cup shredded mozzarella cheese

2 tablespoons grated Parmesan cheese

2 tablespoons minced fresh basil

Hot cooked pasta

✦✦✦

1. Preheat oven to 400°. Pound chicken breasts until thin, about $\frac{1}{4}$ to $\frac{3}{8}$ inch thick. Place in shallow dish and drizzle with olive oil, turning to coat evenly. Sprinkle with garlic powder, basil, salt, and pepper.

2. Preheat contact grill. Grill for 4 to 5 minutes, or until chicken is golden and almost done. Place chicken in 9-x-13-inch baking dish and sprinkle immediately with $\frac{1}{4}$ cup of grated Parmesan cheese. Top with Marinara Sauce.

3. Bake, uncovered, for 10 minutes. Sprinkle with mozzarella cheese and remaining 2 tablespoons of Parmesan cheese. Bake for 5 minutes, or until cheese is melted, chicken is fully cooked, and meat thermometer registers 170°. Sprinkle with minced basil. Serve with hot cooked pasta.

Grilled Chicken Pesto Pasta

Color and flavor packed into an easy, quickly prepared dish. This is one of our favorites.

4 boneless, skinless chicken breast halves

Olive oil

Salt and pepper, to taste

2 tablespoons olive oil

1 green pepper, cut into strips

1 red pepper, cut into strips

½ cup sliced green onions

1 cup pesto (see recipe, page 10)

1 pound penne pasta, cooked according to package directions and drained

⅓ cup shredded Parmesan cheese

1. Preheat contact grill. Pound chicken to ½-inch thickness. Brush chicken breasts with olive oil; salt and pepper to taste.

2. Grill chicken for 5 to 6 minutes, or until meat thermometer registers 170°. Remove chicken and allow to rest for 5 minutes, then cut into strips; set aside.

3. Heat olive oil in large skillet. Cook peppers and green onions until crisp-tender. Add pesto, penne, and chicken strips; toss to combine. Sprinkle with Parmesan cheese.

Makes 6 to 8 servings.
Preparation time: 10 minutes
Grill time: 5 to 6 minutes

Grilled Chicken with Roasted Red Pepper Salsa

TIPS

Roasted Red Pepper Salsa is also great served with tortilla chips!

4 boneless, skinless chicken breasts halves

Olive oil

Salt and freshly ground pepper, to taste

1 can (10 ounces) diced tomatoes and green chiles

1 jar (7 ounces) roasted red peppers, drained and chopped

1 cup frozen corn, thawed

3 green onions, chopped

$\frac{1}{3}$ cup chopped fresh cilantro

✦✦✦

1. Preheat grill. Pound chicken to $\frac{1}{2}$-inch thickness. Brush chicken breasts with olive oil; salt and pepper each side.

2. Grill chicken for 5 to 6 minutes, or until meat thermometer registers 170°. Place chicken on platter and cover with foil; allow to stand for 10 minutes.

3. Combine remaining ingredients and serve with chicken.

Makes 4 servings.
Preparation time: 10 minutes
Grill time: 5 to 6 minutes

Grilled Chicken with White BBQ Sauce

¾ cup mayonnaise

2 tablespoons white wine vinegar

1 teaspoon Creole mustard

½ teaspoon sugar

¼ teaspoon salt

1½ teaspoons coarsely ground pepper

1 clove garlic, minced

1 teaspoon prepared horseradish

6 boneless, skinless chicken breast halves

Olive oil

Salt and pepper, to taste

✦ ✦ ✦

1. Combine mayonnaise, vinegar, mustard, sugar, salt, pepper, garlic, and horseradish; blend well, and set aside.

2. Pound chicken to ½-inch thickness. Brush chicken breasts with olive oil; salt and pepper to taste.

3. Grill chicken for 5 to 6 minutes, or until meat thermometer registers 170°. Serve sauce with chicken.

Makes 6 servings.
Preparation time: 5 minutes
Grill time: 5 to 6 minutes

Grilled Jalapeno Chicken Salad

Chicken salad need not taste like that "same old" dish. Try this new, bold flavor.

4 skinless, boneless chicken breast halves

Olive oil

Salt and pepper, to taste

$\frac{2}{3}$ cup light mayonnaise

1 tablespoon lime juice

$\frac{1}{2}$ cup sliced celery

$\frac{1}{3}$ cup chopped red onion

$\frac{1}{3}$ cup diced red pepper

2 to 3 tablespoons chopped cilantro

$\frac{1}{4}$ cup pickled, sliced jalapenos, drained

Tortilla bowl or tortilla chips

✦✦✦

1. Preheat contact grill. Pound chicken to $\frac{1}{2}$-inch thickness. Brush chicken breasts with olive oil; salt and pepper to taste.

2. Grill chicken for 5 to 6 minutes, or until meat thermometer registers 170°. Allow to rest for 5 minutes, and cut into bite-size cubes; set aside.

3. Combine mayonnaise and lime juice; set dressing mixture aside.

4. Combine celery, red onion, red pepper, cilantro, jalapenos, and chicken. Pour dressing over all, and toss lightly. Spoon into a tortilla bowl or serve with tortilla chips.

Makes 4 to 5 servings.
Preparation time: 5 minutes
Grill time: 5 to 6 minutes

Grilled Lime Chicken with Black-Bean Avocado Salsa

6 boneless, skinless chicken breast halves

$\frac{1}{3}$ cup lime juice

$\frac{1}{3}$ cup chopped cilantro

$\frac{1}{4}$ cup olive oil

1 tablespoon honey

2 garlic cloves, minced

$1\frac{1}{2}$ teaspoons cumin

Salt and pepper

1 can (16 ounces) black beans, rinsed and drained

$\frac{1}{2}$ cup chopped green onions

1 cup frozen corn, thawed

1 red pepper, diced

1 avocado, diced into small cubes

✦✦✦

1. Pound chicken to $\frac{1}{2}$-inch thickness and place in zip-top bag. Combine lime juice, cilantro, olive oil, honey, garlic, cumin, salt, and pepper. Pour $\frac{1}{4}$ cup of lime juice mixture over chicken in bag. Seal, refrigerate, and allow to marinate for 30 minutes.

2. Preheat contact grill. Combine remaining ingredients, and toss with remaining lime juice mixture; set aside.

3. Drain chicken, and discard marinade. Grill chicken for 5 to 6 minutes, or until meat thermometer registers 170°. Serve chicken with black-bean avocado salsa mixture.

Makes 6 servings.
Preparation time: 5 to 10 minutes
Grill time: 5 to 6 minutes

Lemon Chicken with Toasted Bread

4 boneless, skinless chicken breast halves, about 1 pound

$\frac{1}{3}$ cup fresh lemon juice

$\frac{1}{4}$ cup dry white wine

$\frac{1}{2}$ cup olive oil

2 cloves garlic, finely chopped

$\frac{1}{2}$ teaspoon dried thyme leaves

3 tablespoons finely chopped Italian parsley

Salt and pepper, to taste

2 tablespoons olive oil

1$\frac{1}{2}$ to 2 cups fresh bread crumbs

4 tablespoons grated Parmesan cheese

1. Pound chicken until about $\frac{1}{4}$ inch thick; place in large zip-top bag.

2. Combine lemon juice, wine, $\frac{1}{2}$ cup of olive oil, garlic, thyme, parsley, salt, and pepper; blend well and pour over chicken. Refrigerate, and allow to marinate for 30 to 60 minutes.

3. Preheat contact grill. Drain and discard marinade. Grill chicken for 5 to 6 minutes, or until meat thermometer registers 170°. (Cook chicken pieces in batches, if necessary. Remove first batch and keep warm in 300° oven while cooking second batch.)

4. Meanwhile, heat 2 tablespoons of olive oil in skillet. Add bread crumbs and sauté until crisp and golden, stirring often.

5. Place chicken breasts on serving platter. Top each piece with one tablespoon of Parmesan cheese, and sprinkle bread crumbs over all.

Makes 4 servings.
Preparation time: 10 minutes
Grill time: 5 to 6 minutes

Sesame-Crusted Turkey

TIPS

Sesame-Crusted Turkey is especially good with Mushroom Wine Sauce (see recipe, page 300).

For a great salad, slice Sesame-Crusted Turkey into thin strips and arrange on a fresh salad made with torn lettuce, bean sprouts, sliced mushrooms, sliced celery, and other fresh vegetables. Drizzle with vinaigrette dressing.

1½ pounds boneless, skinless turkey breast

2 tablespoons olive oil

1 clove garlic, minced

1 tablespoon soy sauce

1 tablespoon lemon juice

½ teaspoon sesame oil

Freshly ground black pepper, to taste

¼ cup sesame seeds

✴ ✴ ✴

1. Cut turkey into even slices about 1 inch thick; place in zip-top bag. Combine remaining ingredients, except sesame seeds, and pour over turkey. Seal, and squeeze to coat turkey evenly. Refrigerate for 15 to 30 minutes.

2. Preheat contact grill. Drain turkey, and discard marinade. Grill for 6 to 7 minutes, or until turkey is almost done.

3. Sprinkle sesame seeds generously over turkey on grill. Grill for 1 to 2 minutes, or until seeds are golden and meat thermometer registers 170°.

Makes 4 to 6 servings.
Preparation time: 5 minutes
Grill time: 7 to 9 minutes

Southwest BBQ Chicken Salad

TIPS

Any extra dressing may be refrigerated and served on favorite salads, tossed with pasta, or used as a sandwich spread.

1 package (1¼ ounces) sloppy joe seasoning mix

1 tablespoon barbecue sauce

1 tablespoon melted butter

4 boneless, skinless chicken breast halves

1¾ cups ranch dressing

¼ cup barbecue sauce

1 package (1¼ ounces) taco seasoning mix

1 bag (16 ounces) tossed salad mix

1 can (15 ounces) black beans, rinsed and drained

3 green onions, chopped

1 tomato, chopped

1½ cups shredded cheddar, Monterey Jack, or Colby-Jack cheese

1½ cups corn chips

✳✳✳

1. Preheat contact grill. Stir together sloppy joe seasoning mix, 1 tablespoon of barbecue sauce, and melted butter; set aside. Pound chicken to ½-inch thickness. Grill chicken for 4 minutes. Brush chicken with sauce and grill for 1 to 2 minutes, or until meat thermometer registers 170°.

2. Meanwhile, blend together ranch dressing, remaining ¼ cup of barbecue sauce, and taco seasoning; set aside.

3. Toss together salad, black beans, green onions, and tomatoes in a large salad bowl. Cut grilled chicken into thin slices and arrange over salad. Top with cheese and chips. Serve with dressing.

Makes 4 to 6 servings.
Preparation time: 10 minutes
Grill time: 5 to 6 minutes

Southwest Grilled Chicken Salad

TIPS

Check the salad bar at your grocery store. Are there freshly sliced red pepper strips? Maybe black, kidney, or garbanzo beans are there, ready for you to purchase. Buy just the right amount, and forget the waste! It may be just the ticket to add fresh excitement to your salad.

How about giving a new definition to "chef's salad" at your house? When you are the chef, choose what sounds good to you. This version features freshly grilled chicken and a Southwest flavor. We love salads, because they are so easy to adapt to individual yearnings. So first, ask yourself: what sounds good tonight? Go ahead, top it with guacamole, add some leftover cooked corn, omit the beans, or add broccoli florets. It is all good!

4 boneless, skinless chicken breast halves

⅓ cup lime juice

⅓ cup olive oil

2 teaspoons chili powder

1 teaspoon ground cumin

½ teaspoon salt

¼ teaspoon pepper

6 to 7 cups torn lettuce, spinach, or other greens

1 cup canned garbanzo beans or black beans, rinsed and drained

1 small red onion, thinly sliced

1 stalk celery, sliced

1 avocado, peeled, pitted, and sliced

1 tomato, cut into thin wedges

⅓ cup mayonnaise

½ cup shredded Colby-Jack or cheddar cheese

Tortilla chips

✦ ✦ ✦

Makes 4 main-dish servings.
Preparation time: 5 to
10 minutes
Grill time: 5 to 6 minutes

1. Pound chicken breasts to about ½-inch thickness and place in zip-top bag. Combine lime juice, olive oil, chili powder, cumin, salt, and pepper; whisk until well combined.

2. Pour half over chicken; seal bag, and refrigerate for several hours. Cover and refrigerate remaining lime juice-oil mixture.

3. Preheat contact grill. Drain chicken; discard marinade. Grill chicken for 5 to 6 minutes, or until meat thermometer registers 170°. Allow chicken to stand for 5 to 10 minutes.

4. Place lettuce in large salad bowl. Top with beans, red onion, celery, avocado, and tomato. Whisk together reserved lime juice-oil mixture and mayonnaise; toss part of dressing mixture with salad.

5. Thinly slice chicken and arrange over salad. Sprinkle with cheese. Coarsely crush a few tortilla chips, and sprinkle over salad. Serve with remaining dressing mixture.

Teriyaki Chicken and Vegetables

TIPS

Fresh ginger really adds a lot of flavor and is so much better than dried, ground ginger. Fresh ginger is not that expensive; use what you need and place the rest in the freezer until needed again.

The chicken breasts can be pounded, placed in plastic food bags, then frozen. Just set the frozen chicken in the refrigerator the night before you want to cook them.

Chicken breasts are already tender, so the marinade is used to add flavor, not to tenderize the meat. Marinate for just 30 minutes.

Makes 4 servings.
Preparation time: 10 minutes
Grill time: 11 to 14 minutes

Grilled chicken and vegetables are a mainstay in our kitchens. Both of our families enjoy chicken, and we love it because you can add various herbs and seasonings that quickly change the flavor altogether. For example, the fresh flavor of this teriyaki is wonderful, and it is so easy to create. Also feel free to omit the asparagus if not readily available, or substitute green pepper for the red, if you wish.

3 to 4 boneless, skinless chicken breast halves

6 to 8 asparagus spears, trimmed

8 whole fresh mushrooms, halved

1 red pepper, cut into 1-inch strips

1 small zucchini, sliced lengthwise into ½-inch pieces

⅓ cup soy sauce

1 tablespoon minced fresh ginger

2 cloves garlic, minced

2 tablespoons dry sherry

1 tablespoon sugar

1 tablespoon rice wine vinegar

Hot cooked rice

1. Pound chicken to about ½-inch thickness and place in zip-top bag. Place vegetables in a second zip-top bag.

2. Combine remaining ingredients, except rice, and stir well. Measure out and reserve ¼ cup of soy mixture.

3. Pour about half of remaining soy mixture over chicken; seal bag and allow to marinate for 30 minutes. Pour remaining half of soy mixture over vegetables; seal bag, and allow to marinate 30 minutes.

4. Preheat contact grill. Drain chicken and vegetables, and discard marinades. Grill chicken for 5 to 6 minutes, or until meat thermometer registers 170°. Remove and keep warm.

5. Grill about half of the vegetables until vegetables are cooked, yet crisp (about 3 to 4 minutes); remove and keep warm. Repeat with remaining half of vegetables.

6. Arrange chicken and vegetables over rice; drizzle with reserved soy mixture.

Texas Two-Step Pasta with Chicken and Sausage

TIPS

Start water heating as soon as you walk into the kitchen, so everything will be cooked and ready at the same time.

If a spicier dish is preferred, increase cayenne pepper or top pasta with sliced jalapeno peppers.

If desired, substitute spicy or other varieties of smoked sausage for beef smoked sausage.

Makes 4 to 6 servings.
Preparation time: 5 to
   10 minutes
Grill time: 8 to 11 minutes

2 boneless, skinless chicken breast halves

1 tablespoon freshly squeezed lime juice

1/8 teaspoon cayenne pepper

1 tablespoon vegetable oil

1/2 pound smoked beef sausage, cut into 3- to 4-inch pieces

1 package (9 ounces) refrigerated fettuccine

2 tablespoons butter or margarine

2 tablespoons whipped cream cheese

1 jar (16 ounces) salsa

1/4 teaspoon garlic powder

1/8 teaspoon cayenne pepper

1 cup shredded Monterey Jack cheese

1. Pound chicken to 1/2-inch thickness. Place chicken in zip-top bag, and add lime juice, 1/8 teaspoon cayenne, and oil; seal, and shake to coat evenly. Allow to stand for about 5 minutes.

2. Preheat contact grill. Drain chicken. Grill chicken for 5 to 6 minutes, or until meat thermometer registers 170°.

3. Remove chicken from grill; set aside and keep warm. Grill sausages for 3 to 5 minutes, or until hot. Slice chicken and sausages into 1/2- to 1-inch pieces.

4. Meanwhile, cook pasta according to package directions; drain. Place butter and cream cheese in large serving bowl; add hot pasta and toss immediately, so that butter and cream cheese melt and pasta is evenly coated. Add meat to cooked pasta.

5. Pour salsa into microwave-safe container, then stir in garlic powder and cayenne. Cover and microwave on high power (100%) for 1 minute, or until salsa is hot. Pour salsa over pasta and meat; toss gently. Sprinkle with cheese.

FISH AND SEAFOOD

Basil Orange Roughy with Tomatoes

TIPS

If desired, substitute red snapper, cod, or haddock fillets for orange roughy.

¼ cup olive oil

¼ cup minced fresh basil

½ teaspoon salt

½ teaspoon pepper

1 pound fresh, or thawed, orange roughy fillets

1 medium, ripe tomato, sliced about ½ inch thick

2 tablespoons grated Parmesan cheese

✦✦✦

1. Mix together olive oil, basil, salt, and pepper; measure out and reserve half of oil mixture. Brush remaining half of mixture over fish.

2. Preheat grill. Grill fish for about 5 minutes, or until fish flakes easily with a fork. Place on serving platter and keep warm.

3. Grill tomato slices just until hot, for about 30 to 60 seconds. Arrange tomatoes around fish, and immediately sprinkle with Parmesan cheese. Drizzle with remaining olive oil mixture.

Makes 4 servings.
Preparation time: 5 minutes
Grill time: 5½ to 6 minutes

Grilled Fish Tacos

TIPS

The flavor of freshly fried corn tortilla taco shells is well worth the few minutes it takes to make them. Of course, you may substitute packaged crisp taco shells for these if time is short.

Swordfish is a good choice when introducing the flavor of fish to a picky eater. The firm texture and mild flavor make it a favorite. Substitute frozen, thawed swordfish for fresh, if desired.

1 ¼ to 1 ½ pounds swordfish steak, cut about ¾ to 1 inch thick

¼ cup olive oil

2 tablespoons lime juice

2 cloves garlic, minced

¼ teaspoon salt

Dash hot pepper sauce

Cilantro Slaw

3 cups coleslaw mix

¼ cup minced cilantro

2 tablespoons lime juice

2 tablespoons vegetable oil

½ teaspoon sugar

¼ teaspoon ground cumin

¼ teaspoon paprika

Salt to taste

4 to 6 corn tortillas

Vegetable oil

Sour cream, guacamole, or diced tomatoes, optional, for topping

+++

1. Place fish in zip-top bag. Combine olive oil, 2 tablespoons of lime juice, minced garlic, ¼ teaspoon of salt, and hot pepper sauce; pour over fish. Seal bag, refrigerate, and allow to marinate for 30 minutes.

2. Preheat contact grill. Drain fish, and discard marinade. Grill for 3 to 5 minutes, or until fully cooked.

3. Meanwhile, to prepare Cilantro Slaw, mix together coleslaw mix and cilantro. Combine 2 tablespoons of lime juice, vegetable oil, sugar, cumin, and paprika; pour over coleslaw mixture, and toss to coat. Season to taste with salt. (continues on next page)

Makes 4 servings.
Preparation time: 20 minutes
Grill time: 3 to 5 minutes

4. Pour vegetable oil into a small skillet to a depth of about ½ inch. Heat over medium-high heat until oil is hot. With tongs, lower one corn tortilla into oil. Fry just until beginning to crisp.

5. Remove, and place on paper towel to drain. Immediately, using tongs, gently fold in half to form taco shape. Repeat with remaining tortillas.

6. Chop grilled fish. Fill each tortilla shell with diced fish, and top with slaw. Serve with sour cream, guacamole, or chopped tomatoes, if desired.

Grilled Ginger Fish

Nothing beats the aroma of fresh ginger as it grills. This quickly prepared dish will surely call the family to dinner.

¼ cup soy sauce

¼ cup teriyaki sauce

¼ cup brown sugar

1 clove garlic, minced

1 teaspoon grated fresh gingerroot

1 ¼ pounds firm fish steaks, such as halibut or swordfish

✳ ✳ ✳

1. Combine soy sauce, teriyaki, brown sugar, garlic, and gingerroot. Place fish in a glass, 9-×-13-inch dish. Pour soy mixture over fish, cover, refrigerate, and allow to marinate for at least 30 minutes.

2. Preheat contact grill. Drain fish, and discard marinade. Grill for 3 to 5 minutes, or until fully cooked and fish flakes easily.

Makes 4 servings.
Preparation time: 5 minutes
Grill time: 3 to 5 minutes

Grilled Pesto Swordfish

1 ¼ pounds swordfish steaks

3 tablespoons olive oil

2 tablespoons freshly squeezed lemon juice

¼ cup dry white wine

2 cloves garlic, minced

Freshly ground pepper, to taste

¼ cup pesto (see recipe, page 10)

✳ ✳ ✳

1. Place swordfish in shallow dish. Combine olive oil, lemon juice, wine, garlic, and pepper. Pour over swordfish. Coat the steaks with marinade.

2. Preheat contact grill. Drain fish, and discard marinade. Grill for 3 to 5 minutes, or until fully cooked and fish flakes easily.

3. Before removing from grill, drizzle pesto down the center of steaks. Hold grill lid down to heat, but not smash, fish.

Makes 4 servings.
Preparation time: 3 to 5 minutes
Grill time: 3 to 5 minutes

Grilled Seafood Linguine

TIPS

This dish is also excellent when prepared with shrimp, or a combination of shrimp and scallops. Coat seafood with basil-dressing mixture as directed; grill for 1 to 2 minutes, or until shrimp turn pink.

For a wonderful appetizer, prepare shrimp in basil-dressing mixture and grill for 1 to 2 minutes, or just until shrimp turn pink. Serve hot cooked shrimp on toothpicks. Serve remaining basil-dressing mixture as a dipping sauce, if desired.

½ cup fresh basil leaves

½ cup Italian salad dressing

3 cloves garlic

¾ pound bay scallops

1 package (9 ounces) refrigerated linguine

✢ ✢ ✢

1. Combine basil, dressing, and garlic in work bowl of food processor. Pulse to chop evenly.

2. Place scallops in zip-top bag. Pour 2 tablespoons of basil mixture over scallops. Seal, and shake gently to coat evenly.

3. Meanwhile, cook pasta according to package directions; drain.

4. Preheat contact grill. Drain scallops, and discard marinade. Arrange scallops in grill; grill for 1 to 2 minutes, or just until white.

5. Toss together cooked pasta, scallops, and remaining basil-dressing mixture.

Makes 4 servings.
Preparation time: 5 minutes
Grill time: 1 to 2 minutes

Hawaiian Shrimp Salad

TIPS

This recipe is equally good with chicken. Marinate 1 pound of boneless, skinless chicken breasts, pounded to ½-inch thickness, or 1 pound of chicken breast tenders, in marinade as directed. Grill chicken for about 5 minutes, or until fully cooked. Cut into thin strips and add to salad.

To toast almonds, place almonds in a shallow baking pan. Bake at 350° for 10 minutes, or until golden.

Makes 4 to 6 servings.
Preparation time: 30 minutes
Grill time: 4 to 7 minutes
Baking time: 4 minutes

1 pound large shrimp, peeled and deveined
¾ cup pineapple juice
¼ cup soy sauce
3 tablespoons red wine vinegar
2 tablespoons vegetable oil
1 tablespoon honey
1 bag (11 to 12 ounces) salad greens, or 6 to 8 cups torn lettuce
1 red pepper, thinly sliced
1 stalk celery, sliced
16 spears asparagus, trimmed
Olive oil
6 wonton skins
1 tablespoon sesame seeds
⅓ cup sliced almonds, toasted

1. Place shrimp in zip-top bag. Combine pineapple juice, soy sauce, red wine vinegar, and vegetable oil. Measure out ¾ cup of juice mixture and set aside.

2. Pour remaining juice mixture over shrimp; seal bag, and refrigerate for 15 to 20 minutes. Blend honey into reserved juice mixture; cover and refrigerate.

3. Combine salad greens, red pepper slices, and celery in salad bowl.

4. Drain shrimp; discard marinade. Preheat contact grill. Grill shrimp for 2 to 4 minutes, or until shrimp turn pink. Remove from grill, and set aside to cool slightly.

5. Brush asparagus with olive oil; grill for 2 to 3 minutes, or until crisp-tender.

6. Preheat oven to 375°. Cut wonton skins diagonally across to form 12 triangles. Spray a baking sheet with nonstick spray coating. Arrange wonton triangles in single layer on baking sheet; spray wonton skins with nonstick spray coating. Sprinkle with sesame seeds. Bake for about 4 minutes, or until golden and crisp.

7. Arrange shrimp and asparagus on salad. Sprinkle with almonds. Arrange crisp wonton triangles around edge of salad. Drizzle with remaining dressing mixture.

Maple Mustard-Glazed Salmon

Toasting pecans intensifies their flavor. This time, the contact grill not only quickly cooks the salmon, but also toasts the pecans. Add maple syrup, and the result is a delightfully sweet-nutty flavor on the salmon.

2 salmon fillets, about 8 ounces each, cut about 1½ inches thick

Olive oil

Salt and freshly ground pepper, to taste

⅓ cup maple syrup

2 tablespoons Dijon mustard

Dash ground ginger

2 tablespoons chopped pecans

1. Preheat contact grill. Lightly brush both sides of salmon fillets with olive oil, and season to taste with salt and pepper. Grill for 5 minutes.

2. Meanwhile, stir together maple syrup, mustard, and ginger. Brush syrup over tops of salmon. Sprinkle with pecans.

3. Grill for 1 minute, or until fish flakes easily. Serve salmon drizzled with additional syrup.

Makes 4 servings.
Preparation time: 3 to 5 minutes
Grill time: 6 minutes

Sesame Salmon Salad

TIPS

Substitute 6 cups of your favorite lettuce or packaged salad mix for the spinach and bib lettuce in this recipe, if desired.

⅔ cup lemon juice

⅓ cup vegetable oil

1 clove garlic, minced

4 teaspoons sugar

½ teaspoon paprika

½ teaspoon dried basil leaves

½ teaspoon dried tarragon leaves

½ teaspoon seasoned salt

¼ teaspoon pepper

2 to 3 salmon fillets, about 6 to 8 ounces each, cut about 1½ inches thick

1 tablespoon sesame seeds

3 cups torn fresh spinach

3 cups torn bib lettuce

1 carrot, cut into julienne slices

1 cup julienne-sliced zucchini or yellow squash, or combination of the two

1 tomato, cut into wedges

✦✦✦

1. Combine lemon juice, oil, garlic, sugar, paprika, basil, tarragon, salt, and pepper. Measure out 2 tablespoons of lemon juice mixture and brush over salmon. Set remaining lemon juice mixture aside.

2. Preheat contact grill. Grill salmon, skin side down, for 5 minutes; sprinkle tops with sesame seeds. Grill for 1 minute, or until fish flakes easily with a fork and is opaque. Remove from grill, then remove skin from cooked salmon fillets.

3. Meanwhile, combine spinach and lettuce; arrange on serving platter. Top with carrots and zucchini. Place salmon on vegetables; garnish with tomato wedges. Drizzle with remaining lemon juice mixture.

Makes 6 servings.
Preparation time: 10 to 15 minutes
Grill time: 6 minutes

Shrimp and Asparagus over Angel Hair Pasta

TIPS

In the summer, use large tomatoes from the garden. In the winter, use 3 to 4 Roma tomatoes for the best flavor. Chopped tomatoes should measure about 2 cups.

Substitute frozen, thawed shrimp instead of fresh, if desired.

12 ounces large shrimp, shelled and deveined

3 tablespoons olive oil

3 cloves garlic, minced

¼ teaspoon crushed red pepper flakes

12 ounces fresh asparagus, trimmed

1 tablespoon olive oil

8 ounces angel hair pasta

2 large tomatoes, seeded and chopped

2 tablespoons olive oil

1 clove garlic, minced

¼ cup minced fresh basil

Salt and pepper, to taste

Freshly grated Parmesan cheese

1. Place shrimp in a zip-top bag. Drizzle with 3 tablespoons of olive oil, 3 cloves of garlic, and crushed red pepper. Allow to marinate 15 minutes. Place asparagus in another zip-top bag, and drizzle with 1 tablespoon of oil.

2. Meanwhile, cook pasta in boiling salted water, according to package directions; drain.

3. Combine tomatoes, 2 tablespoons of olive oil, 1 clove of garlic, and basil. Season to taste with salt and pepper.

4. Preheat contact grill. Drain shrimp; discard marinade. Grill shrimp for 2 to 3 minutes, or until shrimp turn pink. Remove shrimp from grill and set aside. Add about half of asparagus to grill; grill for 3 minutes. Repeat with remaining half of asparagus. Cut grilled asparagus into 2-inch pieces.

5. Top pasta with tomato mixture, shrimp, and asparagus; toss to combine. Sprinkle with grated Parmesan cheese.

Makes 4 servings.
Preparation time: 30 minutes
Grill time: 8 to 9 minutes

Tilapia with Citrus Ginger Sauce

TIPS

Tilapia is a mild-flavored, freshwater fish. This dish is also good prepared with red snapper or other fish fillets.

The flavor of this citrus ginger sauce is also excellent on grilled chicken breasts.

1 tablespoon vegetable oil

1 teaspoon grated lime zest

2 tablespoons freshly squeezed lime juice

2 teaspoons grated gingerroot

1 pound tilapia fish fillets, cut about ¾ inch thick

2 tablespoons orange marmalade fruit spread

✦✦✦

1. Mix together oil, lime zest, lime juice, and gingerroot in a small, microwave-safe bowl. Measure out about half of oil mixture and brush over fish fillets. Set remaining oil mixture aside.

2. Preheat contact grill. Grill fish for about 3 minutes, or until fish flakes easily with a fork. Place on serving platter.

3. Mix orange marmalade fruit spread with remaining half of lime mixture. Cover, and microwave on high power (100%) for 15 seconds, or just until fruit spread is warm and melted. Brush glaze over fish.

Makes 4 servings.
Preparation time: 5 minutes
Grill time: 3 minutes

Tuna Steaks with Tomato Relish

TIPS

To quickly mince fresh basil, put the washed, trimmed leaves in a 1-cup glass measuring cup. Holding kitchen shears vertically in the cup, quickly snip several times to mince the basil.

We are from the landlocked Midwest and, yes, we both grew up eating canned tuna fish in casseroles and salads. Fresh tuna tastes so much better, and now you can enjoy it often since it is readily available at the fish and seafood counter of most grocery stores.

4 tuna steaks, about 5 to 7 ounces each, cut 1 $\frac{1}{4}$ inches thick

2 tablespoons freshly squeezed lime juice

2 tablespoons olive oil

2 tablespoons minced fresh basil

Tomato Relish

1 clove garlic, minced

3 Roma tomatoes, seeded and diced

3 tablespoons minced fresh basil

1 green onion, chopped

1 tablespoon olive oil

1 tablespoon freshly squeezed lime juice

✦✦✦

1. Place tuna steaks in shallow dish. Combine lime juice, olive oil, and fresh basil, and drizzle over fish. Cover and refrigerate for 30 minutes.

2. Preheat contact grill. Drain fish; discard marinade. Grill for 5 to 6 minutes, or until fish flakes easily with a fork.

3. To prepare relish, combine all ingredients. Top grilled fish steak with relish.

Makes 4 servings.
Preparation time: 10 minutes
Grill time: 5 to 6 minutes

Tuna with Wasabi Mayonnaise

TIPS

Wasabi is also known as Japanese horseradish, and is available in a powder. Look for it in the Asian section of larger grocery stores.

> *Wasabi may be a new flavor for many people, but it is great and adds a bold new flavor to grilled fish.*

1 teaspoon soy sauce

1 teaspoon vegetable oil

2 tuna steaks, about 5 to 7 ounces each, cut 1 ¼ inches thick

¼ cup mayonnaise

¼ teaspoon ground ginger

½ teaspoon wasabi powder

2 to 4 fresh spinach or lettuce leaves

1. Preheat contact grill. Combine soy sauce and oil; brush over each side of tuna. Grill for 5 to 6 minutes, or until fish flakes easily with a fork.

2. Meanwhile, combine mayonnaise, ginger, and wasabi powder. Place grilled fish on spinach leaves. Top each with a dollop of mayonnaise mixture.

Makes 2 servings.
Preparation time: 5 minutes
Grill time: 5 to 6 minutes

VEGETABLES AND SIDE DISHES

343

Grilled Chopped Vegetable Salad

TIPS

If desired, add 1 to 2 cups of cooked, cubed chicken or ham, or 1 pouch (3 ounces) of chunk light tuna to salad just before serving.

1 small sweet onion, sliced ½ inch thick

1 green pepper, seeded and cut in half

1 red pepper, seeded and cut in half

2 firm but ripe tomatoes, sliced about ½ inch thick

2 tablespoons olive oil

2 tablespoons freshly squeezed lemon juice

1 tablespoons olive oil

1 tablespoon balsamic vinegar

1 clove garlic, minced

½ teaspoon dried oregano leaves

Salt and pepper, to taste

8 to 10 pitted kalamata or ripe olives, chopped

Lettuce leaves

✦✦✦

1. Place onion slices, peppers, and tomatoes on a plate. (If necessary, slash peppers so they lay flat.) Brush both sides of vegetables with 2 tablespoons of olive oil, turning to coat evenly.

2. Preheat contact grill. Grill onions and peppers for 10 to 12 minutes, or until somewhat charred; remove to a bowl and cover with aluminum foil. Grill tomatoes for about 1 to 2 minutes, or just until hot.

3. Combine lemon juice, 1 tablespoon of olive oil, vinegar, garlic, oregano, salt, and pepper; whisk to blend well.

4. Coarsely chop onions, peppers, and tomatoes, and place in mixing bowl. Add olives. Drizzle with about half of the dressing mixture, and toss gently to coat.

5. Arrange vegetable mixture on lettuce leaves. Serve with remaining dressing.

Makes 4 servings.
Preparation time: 5 minutes
Grill time: 11 to 14 minutes

Grilled Eggplant with Pasta

TIPS

Salting the eggplant removes some of the bitterness and acid taste. Just salt generously, then rinse and drain.

Substitute penne, ziti, or other pasta for the bow-tie pasta in this recipe.

This dish may not have meat, but it is so rich and flavorful, you can serve it as a main dish and no one will miss the meat. Even if you have not enjoyed eggplant before, give this sensational dish a try.

½ eggplant, peeled and cut into slices ½ to ¾ inch thick
Salt
1 red pepper, cut into fourths
1 medium onion, sliced ½ to ¾ inch thick
10 whole button mushrooms
¼ cup olive oil
12 ounces bow-tie pasta
2 cans (14.5 ounces each) stewed Italian tomatoes, seasoned with basil, garlic, and oregano
2 cloves garlic, minced
¼ cup sliced, pitted ripe olives
1 teaspoon sugar
Salt and pepper, to taste
Freshly grated Romano cheese

✦✦✦

1. Sprinkle eggplant slices generously with salt, then place in colander for 30 minutes. Rinse and drain.

2. Preheat contact grill. Brush eggplant, red pepper, onion slices, and mushrooms with olive oil. Grill eggplant for 3 to 5 minutes, or until golden. Remove grilled eggplant to cutting board. Grill mushrooms for 3 to 5 minutes, or until tender.

3. Grill remaining vegetables for 7 to 10 minutes, or until tender, grilling in batches if necessary. Remove grilled vegetables to cutting board, and allow to cool slightly. Coarsely chop vegetables.

Makes 4 to 6 servings.
Preparation time: 30 to 45 minutes
Grill time: 16 to 20 minutes

(continues on next page)

4. Meanwhile, cook pasta in boiling salted water, according to package directions; drain.

5. Place tomatoes in Dutch oven or large saucepan. Add garlic, olives, and sugar. Heat to a boil; reduce heat and simmer for 5 minutes. Stir in chopped vegetables; simmer for 5 to 10 minutes, or until heated through. Season to taste with salt and pepper.

6. Place cooked pasta into large serving bowl. Add eggplant mixture, and toss to combine. Sprinkle with grated cheese.

Grilled Marinated Portobellos

TIPS

These are great served with grilled steaks.

4 large portobello mushroom caps

½ cup teriyaki marinade

1 teaspoon sesame oil

✦ ✦ ✦

1. Place mushrooms in zip-top bag. Pour remaining ingredients into bag, and let marinate at room temperature for 20 minutes.

2. Preheat contact grill. Remove mushrooms from marinade. Grill for 3 to 5 minutes.

Makes 4 servings.
Preparation time: 3 minutes
Grill time: 3 to 5 minutes

Grilled Parmesan Potatoes

These wonderful potato slices are addictive, and they are a great substitute for French fries.

1 to 2 medium russet potatoes, sliced into $\frac{1}{2}$-inch slices

$\frac{1}{2}$ teaspoon Italian seasoning

$\frac{1}{2}$ teaspoon garlic powder

$\frac{1}{4}$ teaspoon paprika

$\frac{1}{4}$ teaspoon salt

1 tablespoon olive oil

$\frac{1}{4}$ cup shredded Parmesan cheese

✳✳✳

1. Place potatoes in zip-top bag. Combine seasonings and oil; drizzle over potatoes. Seal, and shake to coat well.

2. Preheat contact grill. Arrange potatoes in single layer on grill. Grill potato slices for 12 minutes, or until golden and tender. (If necessary, grill potatoes in batches.) Immediately sprinkle Parmesan cheese over hot grilled potatoes.

Makes 2 servings.
Preparation time: 3 minutes
Grill time: 12 minutes

Grilled Polenta

1 package (16 ounces) refrigerated polenta
Olive oil
Salt and pepper, to taste
Grated Parmesan cheese, optional

✦ ✦ ✦

1. Preheat contact grill. Slice polenta about ½ inch thick. Brush lightly with olive oil.
2. Grill polenta for 3 to 4 minutes, or until golden. Season to taste with salt, pepper, and, if desired, Parmesan cheese.

Makes 4 servings.
Preparation time: 1 to 2 minutes
Grill time: 3 to 4 minutes

Grilled Portobellos on Spring Mix

3 tablespoons extra virgin olive oil

1 shallot, finely chopped

1 tablespoon champagne vinegar

1 teaspoon Dijon mustard

½ teaspoon finely minced fresh thyme

Salt and freshly ground pepper, to taste

4 portobello mushrooms, stems removed and discarded

6 cups spring-mix salad greens

✦✦✦

1. Whisk together olive oil, shallot, vinegar, mustard, thyme, salt, and pepper.

2. Preheat grill. Grill mushrooms for about 3 to 5 minutes, or until very soft and cooked through. Slice mushrooms into strips, and toss with 2 tablespoons of dressing.

3. Toss the spring mix with remaining dressing, and divide among four salad plates. Place mushroom slices in the center of the greens. Serve warm.

Makes 4 servings.
Preparation time: 3 to 5 minutes
Grill time: 3 to 5 minutes

Grilled Vegetable Pasta Salad

TIPS

The woody portion of the asparagus will snap off, leaving the tender shoot.

Substitute a green pepper for the red in this recipe, if you wish. To prepare, cut off stem and top, then cut in half; remove seeds. Wrap half, and return it to the refrigerator to use another time. Cut one half in half, and slash bottom of pepper, if necessary, so the pieces will lay flat on the grill.

Makes 6 to 8 servings.
Preparation time: 10 to
   15 minutes
Grill time: 13 to 18 minutes

Grilling the vegetables makes them sweet and tender—the perfect complement to the pasta and dressing in this salad.

6 to 8 asparagus spears, trimmed
½ red pepper, cut in half
8 to 10 whole button mushrooms
½ small zucchini, sliced lengthwise
2 tablespoons olive oil
1 jar (6.5 ounces) marinated artichoke hearts, drained and chopped
6 ounces bow-tie pasta or macaroni (about 2 cups)
¼ cup olive oil
¼ cup red wine vinegar
1 tablespoon sugar
½ teaspoon dried basil leaves
Salt and pepper, to taste

✳ ✳ ✳

1. Preheat contact grill. Place asparagus, red pepper, mushrooms. and zucchini halves in a zip-top bag. Add olive oil, and shake to coat well.

2. Grill asparagus for about 3 minutes, or until crisp-tender. Grill mushrooms for 3 to 5 minutes, or until tender. Grill remaining vegetables for 7 to 10 minutes, or until crisp-tender. (Cook vegetables in batches, if necessary.)

3. Remove grilled vegetables to cutting board, and allow to cool slightly. Cut asparagus into 1-inch pieces. Coarsely chop remaining vegetables. Place vegetables in large serving bowl, and add artichoke hearts.

(continues on next page)

4. Meanwhile, cook pasta in boiling salted water, according to package directions; drain.

5. Whisk together remaining ingredients; pour over vegetables. Add pasta, and toss to coat well. Cover and refrigerate for several hours, or until well chilled.

Salads!

When dinner sizzles or simmers, often the best accompaniment is a crisp salad. Are you "salad challenged"? Maybe some new, quick tips will transform your salad bowls into salad sensations.

Salad greens go far beyond the head of iceberg lettuce. Experiment, or choose a combination. Salad greens will stay fresher longer if you wash them as soon as you return from the store, then dry the leaves completely. Use a salad spinner, press gently in a towel, or—as Roxanne enjoys—fill a clean pillowcase, go outside, and twirl it around. (Yes, it is a sight, but it works great.) Then, place greens loosely between layers of paper towels, and seal tightly in plastic zip-top bags.

Salads look best if a combination of colors and textures are apparent. Don't just stop with the tomatoes or cucumbers—think of a variety of vegetables, fruits, and nuts. Some items may not be located in the produce isle, but they are still great in salads. Think about water chestnuts, sunflower seeds, dried cranberries, pine nuts, shredded cheese, or roasted peanuts. The list goes on—what about crisp bagel chips instead of croutons, or why not add marinated hot peppers from the Italian section, or quickly thaw frozen peas?

The prepared, packaged salad blends are readily available. Yes, they are convenient, but don't always think of them as finished—they also make a great start to your own customized salad.

Dressings can always change a ho-hum salad into a spectacular one. Try these new salad dressings!

Champagne Vinaigrette

2 tablespoons minced shallots

1 teaspoon Dijon mustard

2 tablespoons champagne vinegar

6 tablespoons olive oil

TIPS

This dressing is also excellent with other types of vinegar. Try white wine, sherry, or red wine vinegar for a change.

✦✦✦

1. Whisk together vinegar, mustard, and shallot. Whisk in olive oil.

Makes about ½ cup.

Gorgonzola Dressing with Walnuts

¼ cup olive oil

1 tablespoon red wine vinegar

Salt and pepper, to taste

4 ounces Gorgonzola cheese, crumbled

½ to 1 cup walnut pieces, toasted

✦ ✦ ✦

1. Whisk together oil, red wine vinegar, salt, and pepper.
2. Drizzle dressing over salad, then sprinkle with Gorgonzola and walnuts.

Makes about ½ cup.

Poppy-Seed Dressing

½ cup vegetable oil

⅓ cup sugar

¼ cup cider vinegar

1 tablespoon poppy seeds

1 ½ teaspoons finely minced onion

¼ teaspoon paprika

¼ teaspoon Worcestershire sauce

Salt and freshly ground pepper, to taste

✳ ✳ ✳

1. Whisk together oil, sugar, and vinegar.
2. Add remaining ingredients, and whisk until combined.

Makes about 1 cup.

Raspberry Vinaigrette

½ cup vegetable oil

¼ cup raspberry vinegar

1 tablespoon honey

½ teaspoon grated orange zest, optional

Salt and pepper, to taste

✦✦✦

1. Combine all ingredients, and shake vigorously.

Makes about ¾ cup.

Sweet and Sour Dressing

TIPS

This dressing is best made a day ahead of time. Chill overnight in the refrigerator to blend flavors; shake well just before serving.

¾ cup vegetable oil

⅓ cup red wine vinegar

⅓ cup sugar

1 teaspoon finely minced onion

1 teaspoon dry mustard

½ teaspoon celery seeds

½ teaspoon salt

¼ teaspoon finely minced garlic

Freshly ground pepper

✦ ✦ ✦

1. Combine all ingredients, and shake vigorously.

Makes about 1¼ cups.

Sesame Asparagus

1 pound fresh asparagus, trimmed

2 tablespoons dry sherry

2 tablespoons olive oil

1 tablespoon soy sauce

1 teaspoon sesame oil

$\frac{1}{4}$ teaspoon salt

$\frac{1}{4}$ teaspoon crushed red pepper flakes

2 tablespoons sesame seeds, toasted

1. Place asparagus in zip-top bag. Combine remaining ingredients, except sesame seeds, and pour over asparagus. Seal bag, and shake gently to coat evenly.

2. Preheat contact grill. Drain asparagus; discard marinade. Arrange asparagus evenly on grill; grill for 3 minutes, or until crisp-tender. Place on serving plate, and sprinkle with sesame seeds.

TIPS

Bend the asparagus gently, and the woody portion at the bottom of each spear will break off.

Toasting the sesame seeds brings out their flavor. Spread the seeds on a baking sheet, and bake in a 350° oven just until seeds are golden brown.

Asparagus–Ham Sandwich Rolls: Prepare Sesame Asparagus, then use the grilled asparagus to make this great recipe. Cut crusts from 8 to 10 slices of white bread, or thickly sliced Texas toast. Use a rolling pin to flatten the bread. Mix together 2 tablespoons of mayonnaise, $\frac{1}{2}$ teaspoon of dried basil leaves, and $\frac{1}{2}$ teaspoon of garlic powder. Spread bread lightly on one side with mayonnaise mixture. Place a thin slice of deli-style ham on bread, then top with 2 grilled asparagus spears. Sprinkle with grated Parmesan cheese. Roll bread around asparagus, and pinch to seal. Brush outside with melted butter. Grill on contact grill for 1 to 2 minutes, or until bread is golden brown and crisp. These sandwiches are great for a lunch treat, or serve as a tasty appetizer.

Makes 4 servings.
Preparation time: 5 minutes
Grill time: 3 minutes

BREAKFAST

Krispy Breakfast Sandwiches

All of those familiar breakfast flavors—ham, eggs, bread, and cereal—are packed into one great-tasting sandwich.

TIPS

You don't have to save these tasty sandwiches for breakfast; serve them anytime.

4 slices firm-textured or country-style white bread

2 teaspoons Dijon mustard

¼ pound thinly sliced deli ham or Canadian bacon

2 slices Swiss or Provolone cheese

1 egg

2 tablespoons milk

Salt and pepper, to taste

1 cup crisp rice cereal

✦ ✦ ✦

1. Preheat contact grill. Spread one side of each bread slice with mustard. Arrange ham and cheese over mustard on 2 slices; top with remaining bread slice, mustard-side down.

2. Whisk together egg, milk, salt, and pepper; pour into a shallow dish. Place crisp rice cereal on a plate.

3. Quickly dip sandwich into egg mixture, coating both sides evenly. Gently press sandwich into cereal, again coating each side evenly. Repeat with remaining sandwich.

4. Grill sandwiches on contact grill for 2 to 4 minutes, or until outside is golden brown and cheese is melted.

Makes 2 servings.
Preparation time: 5 to 10 minutes
Grill time: 2 to 4 minutes

Orange French Toast

For a quick breakfast, and make-ahead convenience, place the dipped bread slices on a tray lined with wax paper. Freeze just until bread is firm. Remove frozen bread from tray, place in zip-top bags, and return to freezer. When ready to cook, just preheat the grill, then grill frozen slices for 1 to 2 minutes; no need to thaw first.

This golden French toast is great right off the grill. Try the frozen tip and keep some ready to grill for a really quick, hot breakfast.

¾ cup orange juice

3 eggs, lightly beaten

1 tablespoon confectioners' sugar

½ teaspoon ground nutmeg

Dash salt

1 teaspoon vanilla

10 to 12 slices baguette, each cut about 1½ inches thick

Butter, syrup, confectioners' sugar, or orange marmalade

✦✦✦

1. Preheat contact grill. Whisk together orange juice, eggs, sugar, nutmeg, and salt. Quickly dip bread into juice mixture, coating evenly on both sides.

2. Spray grill with butter-flavored nonstick spray coating. Grill bread for about 1 to 2 minutes, or until golden. (Watch closely, as it will brown quickly.)

3. Serve warm with butter and syrup, dust with confectioners' sugar, or top with orange marmalade.

Makes 5 to 6 servings.
Preparation time: 3 to
  5 minutes
Grill time: 1 to 2 minutes

DESSERTS

Apple-Walnut Quesadillas

> *While quesadillas are a favorite appetizer, they are not often thought of for dessert. This recipe may just change that!*

TIPS

Toasting the walnuts intensifies their flavor. To toast, spread walnuts on a baking sheet and bake at 350° for about 5 to 7 minutes, or until golden brown.

When grilling these quesadillas, place the filled quesadilla with the open edges of the tortilla toward the hinges of the contact grill. This way, the folded edge will be toward the lower, open end of the contact grill, and it will retain more of the butter and sugar. This is especially true if the contact grill you are using is designed at an angle so fats can drip off.

Makes 4 servings.
Preparation time: 10 to 15 minutes
Grill time: 8 minutes

3 medium tart cooking apples, such as Granny Smith, peeled, cored, and thinly sliced

$\frac{1}{4}$ cup brown sugar

1 teaspoon ground cinnamon

1 tablespoon water

4 flour tortillas, about 8 inches in diameter

$\frac{1}{4}$ cup melted butter

$1\frac{1}{2}$ teaspoons sugar

$\frac{1}{2}$ teaspoon ground cinnamon

$\frac{1}{4}$ cup cream cheese, softened

$\frac{1}{2}$ cup chopped walnuts, toasted

6 tablespoons caramel sauce, warmed

$\frac{1}{4}$ cup chopped walnuts, toasted

Sweetened whipped cream

✦ ✦ ✦

1. Combine apple slices, brown sugar, 1 teaspoon of cinnamon, and water in a small saucepan. Cook over low heat, stirring occasionally, for 10 to 15 minutes or until apples are tender.

2. Preheat contact grill. Brush one side of each tortilla with melted butter. Combine sugar and $\frac{1}{2}$ teaspoon of cinnamon. Sprinkle $\frac{1}{2}$ teaspoon of cinnamon-sugar mixture over buttered side of tortilla. Spread 1 tablespoon of cream cheese evenly over the other side of the tortilla.

Want a quick alternative to the cooked apples? Substitute canned pie filling, in your favorite flavor, for the fresh apples. Then, feel free to adjust the type of nuts to complement the fruit. For example, you could add slivered almonds to a cherry pie filling, or you might choose pecans to go with peach filling.

3. Spoon about $\frac{1}{4}$ of apple mixture over cream cheese, then sprinkle with about 2 tablespoons of chopped walnuts. Fold tortilla in half. Grill for about 2 minutes, or until golden brown and crisp.

4. To serve, place a toasted quesadilla on serving plate. Drizzle with warm caramel sauce, and garnish with 1 tablespoon of chopped walnuts. Dollop with whipped cream.

Grilled Alaskas

TIPS

Assemble the cake and ice cream sandwiches at least one day ahead, or up to a week ahead. Just wrap tightly in plastic wrap and keep solidly frozen until ready to use.

For an added touch, puddle fruit purée next to each dessert. To make a raspberry purée, place 1 package (12 ounces) of frozen unsweetened raspberries, thawed, in the work bowl of a food processor. Add $\frac{1}{3}$ to $\frac{1}{2}$ cup of sugar; process until smooth. Pour through a sieve to remove seeds. Makes about $\frac{1}{2}$ cup of purée.

Sweetened whipped cream is a snap to make. Beat 1 cup of heavy whipping cream until frothy. Gradually add 2 tablespoons of confectioners' sugar, and continue beating until stiff peaks form. Of course, you can use frozen, nondairy whipped topping, thawed, if time is limited.

Makes 4 servings.
Preparation time: 5 minutes
Grill time: 1 to 2 minutes

Baked Alaska—that classic dessert of a baked cake topped with ice cream and meringue—takes a new approach on the contact grill. Be sure to assemble the ice-cream-filled cake "sandwiches" at least one day before, so they are solidly frozen.

1 cup ice cream, any flavor, softened
8 slices pound cake, each cut about $\frac{3}{4}$ inch thick
Sweetened whipped cream
$\frac{3}{4}$ cup chocolate sauce

✶ ✶ ✶

1. Spread ice cream evenly over 4 slices of pound cake, mounding slightly in the center. Top with a second slice of pound cake. (Place second slice of cake on gently, so it doesn't squeeze the ice cream off the cake.) Wrap each sandwich tightly in plastic wrap, and freeze until quite solid.

2. Preheat contact grill. Unwrap sandwich, and grill for 1 to 2 minutes, or just until pound cake is lightly toasted. (Grill in batches to avoid overcrowding; keep frozen until ready to grill.)

3. Place on individual serving plates. Dollop each sandwich with sweetened whipped cream. Drizzle with chocolate sauce. Serve immediately.

Grilled Peach Sundaes

Grilled peaches, warm and slightly caramelized from the grill, combined with fresh peach ice cream tastes heavenly.

4 ripe but firm fresh peaches, halved

2 tablespoons almond-flavored oil or vegetable oil

⅓ cup sugar

1 teaspoon cinnamon

4 scoops peach- or vanilla-flavored ice cream

½ cup caramel sauce

✦✦✦

1. Remove pits from peaches. Brush cut surface of peaches lightly with oil. Combine sugar and cinnamon; dip cut surfaces of peaches into cinnamon-sugar mixture.

2. Preheat contact grill. Place peaches cut-side down; close top of grill lightly. Grill for 1 to 3 minutes, or just until sugar is caramelized.

3. Place peaches in four individual serving dishes. Top warm peaches with a scoop of ice cream, then drizzle with caramel sauce.

Makes 4 servings.
Preparation time: 5 minutes
Grill time: 1 to 3 minutes

Grilled Pineapple with Ice Cream

¼ cup brown sugar

2 tablespoons butter

8 slices ripe fresh pineapple, about ½ inch thick

2 cups vanilla ice cream

✦ ✦ ✦

1. In a small saucepan, cook the brown sugar and butter over low heat until sugar dissolves and bubbles. Keep sauce warm until ready to serve.

2. Preheat grill. Grill pineapple slices for 3 to 5 minutes, or until golden and softened.

3. Place 2 pineapple slices on dessert plates, top with one scoop ice cream, and drizzle with brown sugar sauce.

Makes 4 servings.
Preparation time: 5 minutes
Grill time: 3 to 5 minutes

Grilled Pound Cake with Berries

TIPS

Substitute fresh raspberries or blueberries for strawberries, if desired. Substitute frozen, thawed fruit for fresh, if necessary to make this a year-round treat. If preparing a purée with raspberries, purée until smooth, then pour purée through a sieve to remove seeds.

To toast sliced almonds, preheat oven to 350°. Spread almonds in single layer on baking sheet. Bake for 5 to 7 minutes, or until golden.

2 cups sliced fresh strawberries

2 to 4 tablespoons sugar

1 tablespoon Grand Marnier or Chambord, optional

6 slices pound cake, each about ½ to ¾ inch thick

6 scoops vanilla ice cream

Sweetened whipped cream

½ cup sliced almonds, toasted

1. Preheat contact grill. Combine sliced strawberries with sugar in food processor; pulse to chop and combine into a chunky puree. Stir in liqueur, if desired. Set aside.

2. Grill pound cake for 1 to 2 minutes, or just until golden brown. Immediately place each slice of toasted pound cake in each serving dish.

3. Top warm cake with a scoop of ice cream, then spoon fruit over ice cream. Dollop with whipped cream, and garnish with sliced almonds. Serve immediately.

Makes 6 servings.
Preparation time: 5 minutes
Grill time: 1 to 2 minutes

Strawberry Dessert Sandwiches

TIPS

Keep frozen pound cake on hand. Thaw it just enough to slice it, then wrap the slices in plastic wrap and return to freezer. That way, you can thaw just the number of slices you might need.

This is a great spur-of-the-moment dessert.

4 slices pound cake, each about $\frac{1}{2}$ to $\frac{3}{4}$ inch thick

$\frac{1}{4}$ cup whipped strawberry-flavored cream cheese, at room temperature

4 strawberries, sliced

2 tablespoons semi-sweet chocolate chips

Chocolate sauce

Sweetened whipped cream

✦ ✦ ✦

1. Spread one side of 2 slices of pound cake with cream cheese. Top cream cheese with sliced strawberries and chocolate chips, then top with second piece of pound cake to make a sandwich.

2. Preheat contact grill. Grill pound-cake sandwiches for 1 to 2 minutes, or just until cake is golden.

3. Place each on a serving plate. Drizzle with chocolate sauce and garnish with a dollop of sweetened whipped cream.

Makes 2 servings.
Preparation time: 3 minutes
Grill time: 1 to 2 minutes

Toasted Chocolate

This is a grand "after-school" snack on a chilly day. One bite, and you will be reminded of European delights!

4 slices fine-textured artisan bread

Butter, softened

2 ounces semi-sweet chocolate pieces

✦ ✦ ✦

1. Butter one side of each slice of bread. Cover 2 slices of bread, on non-buttered sides, with chocolate. Top with second slice of bread, buttered-side out.

2. Preheat contact grill. Grill for 3 to 5 minutes, or until chocolate is melted. Cool slightly and cut sandwiches in half.

Makes 2 servings.
Preparation time: 3 minutes
Grill time: 3 to 5 minutes

Tropical Glazed Banana Sundaes

TIPS

Toasting the almonds and coconut intensifies the flavor. To toast either, spread in a single layer on a baking sheet. Bake at 350° just until golden brown, about 5 to 10 minutes.

Grilled and glazed bananas, with the remaining warm syrup, are also great on pancakes for breakfast.

¼ cup butter

2 tablespoons sugar

2 tablespoons brown sugar

2 tablespoons lemon juice

1 teaspoon cinnamon

4 firm, but ripe, bananas

4 scoops ice cream

¼ cup sliced almonds, toasted

¼ cup flaked coconut, toasted

✦ ✦ ✦

1. Melt butter in small saucepan over low heat. Stir in sugar, brown sugar, lemon juice, and cinnamon, stirring until sugar is dissolved.

2. Preheat contact grill. Peel bananas and, immediately, lightly brush glaze evenly over bananas.

3. Grill for about 2 minutes, or until bananas are hot. Remove from grill and cut diagonally into fourths.

4. Place a scoop of ice cream in 4 individual serving dishes. Top each with grilled bananas. Drizzle remaining glaze on top of ice cream. Sprinkle with toasted almonds and coconut.

Makes 4 servings.
Preparation time: 5 minutes
Grill time: 2 minutes

Index